DOWN

WIND

A PEOPLE'S HISTORY OF THE

NUCLEAR WEST   *Sarah Alisabeth Fox*

for my dear Barb —
thank you for
your encouragement
& support!

UNIVERSITY OF NEBRASKA PRESS | LINCOLN AND LONDON

Library of Congress Cataloging-
in-Publication Data
Fox, Sarah Alisabeth, 1981–
Downwind: a people's history of the
nuclear West / Sarah Alisabeth Fox.
pages   cm
Includes bibliographical
references and index.
ISBN 978-0-8032-5537-1 (cloth: alk. paper)
ISBN 978-0-8032-6950-7 (epub)
ISBN 978-0-8032-6951-4 (mobi)
ISBN 978-0-8032-6949-1 (pdf)
1. Nuclear weapons testing victims—West
(U.S.) 2. Nuclear weapons—Testing—West
(U.S.)—History. 3. Nuclear weapons—
Testing—Health aspects. 4. Radioactive
fallout—Health aspects—West (U.S.)
5. Radiation—Health aspects—West (U.S.)
6. Uranium mines and mining—Health
aspects—West (U.S.) 7. Uranium miners—
Health and hygiene—West (U.S.) I. Title.
II. Title: People's history of the nuclear West.
RA569.F69 2014
363.7380978—dc23
2014023111

Set in Scala by Lindsey Auten.
Designed by N. Putens.

To
Michelle and Irma Thomas,
Claudia and Bethany Peterson,
Mary Dickson,
Preston Jay Truman and Monte Bright,
Ivan Sidney,
Timothy Benally,
Phil Harrison,
Manuel Pino,
Robin, Kaaren, and Roger Ray, Christena Peterson,
Corbin Harney,
Loa, Blaine, and Sybil Johnson,
Dave Timothy,
Janet Seegmiller,
McRae and Lillian Bulloch,
Quentin and Gwendolyn Nisson,
Sheldon and LaVerna Johnson,
Willard Peterson,
Tona and John Henderson,
their families,

and the others who are too many to name.

CONTENTS

ILLUSTRATIONS

ACKNOWLEDGMENTS

The individuals who shared their painful and important stories with me made this project possible. In gratitude to them, and in recognition of the amount of work that needs to be done on behalf of the people and lands devastated by nuclear testing and uranium extraction, I am donating a portion of the proceeds from the sale of this book to HEAL (Healthy Environment Alliance) Utah and the Southwest Research and Information Center to support their continuing work on these issues.

Shortly after I moved to Utah to pursue a graduate degree in history, I reread Terry Tempest Williams's book *Refuge* and realized with a start that I had arrived in downwinder country. Her words of encouragement in a November 2004 book-signing line gave me the courage to go forward in the early months of my research and still do, ten years later. Carole Gallagher's extensive oral history work in *American Ground Zero* introduced me to the stories of many citizens who have since passed on, providing a critical link to several generations of expert witnesses. I am indebted to her years of fieldwork.

Peter Eichstaedt, Timothy Benally, Phil Harrison, and Doug Brugge's interviews with indigenous uranium workers were similarly invaluable to this project. Financial assistance from Utah State University's History Department, the *Western Historical Quarterly*, and Morris and Frances Fox was crucial at the outset of my research. David Rich Lewis's willingness to support my interdisciplinary work brought me to Utah in the first place, and his mentorship, advocacy, good humor, and honest opinions have shaped me as a scholar and a citizen. Advisers Steve Siporin, Gary Madsen, and Chris Conte offered valuable insights during early research and writing stages. Archivists Robert Parson (Utah State University) and Janet Seegmiller (Southern Utah University) helped unearth important material.

Matthew Bokovoy skillfully and patiently shepherded my manuscript through several years of revisions, and Sabrina Stellrecht and Vicki Chamlee shared their incisive editorial oversight in the final stages. I am indebted to them and all the other experts behind the scenes at the University of Nebraska Press for their good work.

I extend my heartfelt gratitude to Ed Singer and Sonja Horoshko, who granted permission for Ed's extraordinary painting *Dear Downwinder* to grace the cover of this work. By placing the human subject in the foreground and illustrating the terror of exposure to both the Yellow Monster of uranium and the looming mushroom cloud of the test site itself, Ed has communicated the experience of life downwind in ways my words cannot. I also owe thanks to Roxanne Dunbar-Ortiz for helping my book to find a good home, cartographer Erin Greb for her elegant maps, graphic designer Emma Fox for painstakingly restoring and preparing illustration materials, Alicia Engstrom for assiduously constructing the book index, and Andrea Fuentes-Diaz for her stellar photography. And while none of them were involved directly in this project, Linda Ellis, Becky Firth, Linda Larrabee, Kurt Dunbar, Bob Keller, Ona Siporin, Sam Schrager, and David Marr have all lent me extraordinary guidance in developing my toolkit as a researcher and a writer.

I owe much to my parents. My father, Howard Fox, taught me to honor the stories and struggles of working people, showed me by example how to listen compassionately, and cherished my writing in all its forms. My mother and longtime mentor, historian Theresa Trebon, shared her wisdom on method and insight into material, sent innumerable references my way, and accompanied me on major fieldwork trips, poring over microfilm, assisting me in interview preparation, and joining me as I wandered cemeteries and learned the land. It is an honor to call myself her colleague.

Four years before this project came to fruition, my son, Callum, arrived on the scene, imbuing my work with new meaning and a previously unknown degree of challenge. Dozens of family members and friends cared for him as I waded through revisions. I hope they know this book would not have been possible without their loving care. From an early age Callum graced this project with his patience and sweetness. As it goes to press, he is finally tall enough to reach the printer without a stool and to retrieve pages for me, and he continues to forgive me for the amount of attention "Mama's book" demands.

Finally, without the erstwhile support and encouragement of Ryan Reilly—my dearest friend, my husband, and the father of our son—this book would still be a stack of research in a file cabinet. He held me accountable to my vision, talked out innumerable logistics, paid bills, updated my recording equipment, baked me cookies, drove cross country, and set up writing retreats. Whenever my courage failed, he pushed me to go further, and this book owes its very existence to his tenacious support. My gratitude to him is as deep and wide as our Pacific.

DOWNWIND

{1} LIVING UNDER THE CLOUD

I was outside with my brother and I saw this big red ball come over up over the horizon and I thought it was a flying saucer so I ran to the house to tell my mother.

Claudia Peterson, St. George, Utah[1]

By the time five-year-old Claudia returned to her swing set, a strangely colored cloud was all that remained of her flying saucer. Years later, she learned the apparition she had seen in the sky was not a UFO but the mushroom cloud of a nuclear explosion. Her childhood home in southern Utah was about a hundred miles east of the Nevada Test Site (known today as the Nevada National Security Site), one of the most heavily utilized nuclear weapons testing areas in the world. From 1951 to 1992, regular nuclear explosions rattled the region, scattering untold quantities of radioactive isotopes into the air, soil, and water, permeating the food chain downwind of the test site. Many families in the region either kept livestock and gardens or bought meat, milk, and produce from their neighbors, unwittingly gathering radiological contamination in their backyards and placing it on their dinner tables. By the age of thirty-five, Claudia Peterson had survived cancer and had buried her six-year-old daughter, her sister, her father, and numerous neighbors and friends, all of whom had succumbed to diseases potentially caused by radiation exposure.[2]

There are thousands of stories like Claudia's. For many, there was no flash on the horizon; instead, one year a uranium mine or mill opened near their homes, offering good wages to workers who had few options for earning an income. The vast majority of the nation's domestic uranium reserves, a crucial component in the development of nuclear weapons, existed southeast of the test site in the rural desert land of the Four Corners region. A significant portion of those reserves lay beneath the soil and rocks of Indian reservations, predominantly those of the Navajo and Pueblo peoples. While some indigenous residents of the region expressed serious qualms about the spiritual and ecological risks of dynamiting and tunneling into the earth, the prospect of wage labor close to home lured many into the industry. "We were blessed, we thought," Navajo miner George Tutt recalled. "Railroad jobs were available only far off like Denver . . . but for mining, one can just walk to it in the canyon. We thought we were very fortunate, but we were not told, 'later on, this will affect you in this way.' . . . We called it 'good work.'"[3]

Milled and refined, the rock the Navajo called *leetso* — "yellow brown" or "yellow dirt" — became the volatile material at the heart of the bombs detonated in the desert a few hundred miles to the west. As the years passed, the men who had made their living in the mines and mills began to succumb to lingering illnesses, and many did so at relatively young ages. Stories cohered in small communities about entire generations of men who began coughing up blood, struggled to draw breath, and died young. Many left behind widows bound by hardship, living in poverty and caring for multiple children and family livestock holdings. Women such as Betty Jo Yazzie, Grace Tuni, and Rose Benally noticed how other men who had worked with their husbands in the mines and mills died in the same way their husbands did. They wove stories around their losses, tying together the maladies of the laborers with the sicknesses of animals and families who lived downstream from the mines and downwind from the uranium tailings piles. Like thousands of their neighbors, Claudia Peterson and Betty Jo Yazzie came to understand their losses were more than local or personal tragedies. In their search for answers they learned they had grown up and raised families on the toxic staging grounds of the Cold War.[4]

On 16 July 1945, six years before the atomic bombs began raining down on the Nevada desert, the first atomic bomb in history was detonated in White Sands, New Mexico. A few weeks later the second and third atomic bombs in history followed, detonated over the cities of Hiroshima and Nagasaki, Japan. While hundreds of thousands of Japanese citizens succumbed to injuries and radiation sickness, the U.S. government falsely described the targeted cities as military bases and the bombs as nothing special, just larger, more efficient conventional weapons. Newspapers and magazines across the country faithfully repeated these descriptions and opinions as fact. "The Japanese began the war from the air at Pearl Harbor," read President Harry Truman's 6 August statement announcing the bombing. "They have been repaid many fold." Military leaders such as Maj. Gen. Leslie Groves adamantly denied reports of radiation sickness in Japan and insisted the bombs had hastened the end of a long and bloody war. The voices of dissenting journalists, scientists, clergymen, and citizens were censored and drowned out. A Gallup poll taken in August 1945 concluded 85 percent of Americans approved of the atomic bombings of Japan. When the dangers of radiation exposure could not be denied any longer, General Groves told a Senate special committee that doctors had described radiation sickness as "a very pleasant way to die."[5]

Cracks did emerge in the state-sponsored narrative, particularly after journalist John Hersey published "Hiroshima" in the August

1946 issue of the *New Yorker*. Based on first-person interviews and research conducted after the city was bombed, Hersey's story was the only content of the magazine that month, and it was subsequently read aloud on ABC radio and reprinted in newspapers nationwide. Americans gained their first glimpse into an atomic bombing from the perspective of the bombed. The vivid descriptions of radiation sickness, poison dust and rain, and the relatable characters—for example, a mother, a priest, a young woman—humanized the Japanese, who for so long had been only a vilified stereotype, and firmly established the gruesome and long-lasting damage of the new superweapon.

While it fundamentally altered many Americans' opinion of atomic weaponry, Hersey's story could not change the circumstances in which the nation's leaders found themselves. Having built and used an atomic weapon in war, the United States could not expect to hold a monopoly on the weapon for long. Russia was actively pursuing the atomic bomb, and with tensions simmering between the two nations, American military leaders concluded that establishing a larger, more sophisticated nuclear arsenal was their only recourse. An aggressive, heavily propagandized program of weapons development emerged. With Russia's first successful nuclear test in August 1949, competition between the two nations ignited a global arms race that would last for the next four decades.

A central component of this race was the testing and refinement of newly designed weapons. Military leaders and scientists wanted to know how large the explosions would be; how the bombs would react to being dropped from planes and fired from cannons; how troops, buildings, ships, weather patterns, and military uniforms would be affected by the explosions; and how significant the ensuing radiological contamination would be. Some tests were designed to illustrate civil defense tactics for surviving a nuclear blast, while others were conducted to convince the public that nuclear weapons could be used for practical tasks such as excavating canals, reservoirs, and harbors. Gwendolyn Nisson, who raised her children close enough to the test site that it rattled her windows, later compared the men running the tests to "a bunch of little boys with firecrackers."[6]

In the immediate wake of World War II, amid celebrations of the supposed return to peacetime, selling the war-weary public on a costly and dangerous weapons development program took some marketing. The public needed to be afraid enough of enemy weapons that they would throw their support behind the development of American nuclear technology but not so afraid that they would protest weapons testing or support a ban on nuclear weapons altogether. To convince the public of the need for the testing program, "U.S. military officials began to spin visions of future conflicts before World War II was even over," writes historian Patrick Sharp. "Their stories focused on what technologies and strategies the United States needed to develop to win the next major war." Sharp calls these stories strategic fictions. Based on the presumption that massive military strength could deter an enemy attack, strategic fictions garnered their legitimacy from their authors, military men such as head of U.S. Army Air Forces Gen. Henry H. "Hap" Arnold.[7]

On 19 November 1945, an article titled "The 36-Hour War" appeared in *Life* magazine that detailed General Arnold's vision of a future nuclear attack on the United States. While the article claimed the United States triumphed in the imaginary conflict, wiping out the enemy with a nuclear attack, victory exacted a high cost: "Some 40,000,000 people have been killed and all cities of more than 50,000 population have been leveled." "The 36-Hour War" came across as more prophetic than theoretical, with Arnold's military credentials helping to blur the lines between fact and fiction. Uncritically advancing Arnold's claim that the nation's security rested on an ability to "take immediate offensive action with overwhelming force," the *Life* piece impressed readers across the nation—including those living near what would soon become the Nevada Test Site—with the dire need to develop new weapons technology.[8]

We are accustomed to calling the ensuing era the Cold War, but it was as hot as they come. Writer Rebecca Solnit describes nuclear weapons testing as "a way of making war by display and displacement, as some cultures and species do—demonstrating their ability to attack rather than actually doing so." The explosions Claudia Peterson witnessed from her childhood home were every bit as real as the ones that had occurred over Hiroshima and Nagasaki, and most of them were many times more powerful. Americans were told that having a large nuclear arsenal would protect national security, that it was their patriotic duty to build bomb shelters in their homes to protect them from the Russians, and that the tests in Nevada were controlled experiments.[9]

In actuality, few aspects of the tests could really be controlled. No one knew precisely how large the explosion of any given bomb would be or if it would explode at all. Tests only went forward when the wind was blowing east—away from densely populated California and over Utah, Idaho, Arizona, and the states that lay beyond; directly over Claudia Peterson's backyard; and frequently over the homes of the families of the very men who had unearthed the uranium to build the bombs.

In a fundamental conflict of interest, the officials overseeing the uranium industry and the construction and detonation of the largest bombs in history represented the very agency that was also responsible for monitoring contamination and protecting citizen health. Created by the 1946 Atomic Energy Act, the Atomic Energy Commission (AEC) ostensibly put control of nuclear technology in civilian hands. In actuality, the agency worked intimately with the military on weapons development and answered to almost no one, having been given carte blanche to act with complete secrecy in the pursuit of all matters nuclear. Well into the 1970s, the AEC conducted nuclear experiments at the risk of the health of their own citizens, all the while quietly monitoring the results with impunity.

AEC representatives told those citizens who lived immediately downwind and who participated in the uranium industry that their

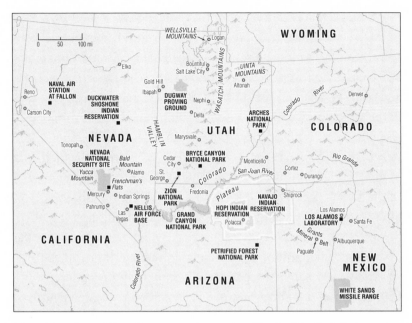

Map 1. Test site and Four Corners uranium region. Map by Erin Greb Cartography.

cooperation, support, and labor were important contributions to national security and that the test site and the uranium industry posed no danger to them or their families. More subtly, citizens in this region were made to understand that asking too many questions about safety was unpatriotic. The national media regularly communicated and underscored this message with propagandistic themes, affirming that threats to Americans' liberty and safety were many and imminent and that every government action—censorship, weapons testing, or suppression of dissenting opinions—was taken to protect the nation. For the first few years, most in the downwind and uranium-producing region wholeheartedly supported the AEC's endeavors.

Citizens who lived in proximity to the Nevada Test Site and the adjacent uranium industry during this era shared several fundamental characteristics. Many were actively involved in farming or ranching or came from families that had produced their own food in the past. Fishing, hunting, and gathering were popular as both recreational activities and sources of sustenance. Strong work ethics and cooperation were prized in these communities, where success and comfort depended on great effort and collaboration at key times in the agricultural cycle. Some communities had been continuously occupied by Native residents for hundreds of years, while others dated only to the nineteenth century, when they were settled by members of the Church of Jesus Christ of Latter-Day Saints (also referred to as the LDS or Mormon Church). Despite a long-standing tradition of antipathy toward the federal government among LDS members in particular, and across the American West in general, most in the downwind and uranium-producing region were self-identified patriots who prized national security and enthusiastically supported the U.S. government's efforts in the Cold War. These citizens tended to live in accordance with social norms, and they trusted that doing so would protect their families and their way of life.

When they learned their government had actively concealed the risks of radiological exposure from nuclear weapons development and uranium extraction, had knowingly exposed them to dangerous pollutants, and had contaminated their food supply, many of these citizens experienced a profound sense of betrayal. Not only had many lost their health and their loved ones, but also they felt deceived and abandoned by the government to which they had given their unquestioning loyalty. Farmers and ranchers who had labored with pride to raise milk, meat, and crops realized they had been exposed to radiation while doing so and had most likely exposed the consumers of their products to that same radiation. Some experienced significant economic losses when their livestock succumbed to radiation poisoning, and as family farming began to wane in the United States in subsequent decades, many in downwind regions blamed nuclear testing or the uranium industry. They saw their losses—of their family members, of their health, of their economic solvency, of their trust in their leaders—as connected, and many felt their way of life had been destroyed. Many began to identify themselves as "downwinders" or "uranium-affected people" or both and to question what they had taken for granted: the safety of food, the word of trusted leaders, and the meaning of "national security."

As modern-day observers, our first question upon learning of this disturbing chapter in U.S. history is invariably a question of scope. How many bombs? How much uranium? How much sickness? The Federation of American Scientists tallies 2,046 nuclear weapons tests conducted worldwide between 1945 and 2002. Of those tests, 504 nuclear weapons were detonated in the atmosphere, with the United States exploding 100 of them. Others were tested under the surface of the ocean. When atmospheric testing ceased in 1962, the United States subsequently tested more than 900 nuclear weapons underground. More than 225 million tons of uranium were extracted in the United States alone between 1950 and 1989.[10]

The health ramifications of radiological exposure are many and varied. For those exposed to significant levels of radiation, documented immediate aftereffects include "beta burns, loss of hair, depressed red cell and leukocyte counts, flulike symptoms, nausea, fingernail discoloration, radioactivity in the urine, and changes at the cellular level in blood and bone marrow." Studies conducted on residents of the Marshall Islands, who were exposed to radiation during U.S. nuclear tests in the Pacific Ocean, revealed long-term effects that included "immune-deficiency diseases, metabolic disorders (diabetes), growth impairment in children, cancers, leukemia, premature aging (dental decay, cataracts, degenerative osteoarthritis), and a host of reproductive problems including miscarriages, congenital birth defects, and sterility."[11]

How many people were affected? Estimates vary widely, and all are controversial. In 1969 Dr. John Gofman, founder and director of the Atomic Energy Commission's biomedical research division at Lawrence Livermore Laboratory, estimated that "the risk of cancer [from radiation exposure] had been underestimated at least twenty times" and "predicted . . . that if everybody got the [AEC's] 'permitted' dose there would be something like 16,000 to 32,000 extra cancer deaths per year in the United States." A number of Gofman's colleagues, many of whom were on the AEC and nuclear industry payrolls, dismissed his figures, and he later lost his research funding. That same year, Dr. Ernest Sternglass suggested that the deaths of as many as 400,000 infants could be connected to fallout-related malaises such as low birth weight and decreased immune function. Six years prior, Sternglass had been a go-to expert on the connection between prenatal radiation exposure and increased rates of infant mortality and childhood leukemia, and his research helped generate support for the 1963 Limited Test Ban Treaty, which drove nuclear testing underground. As Sternglass's later research pointed to higher and higher infant mortality rates, other scientists roundly criticized his methodology and results. While he did not support Sternglass's conclusions, well-respected nuclear physicist Freeman Dyson wrote in 1969, "The evidence is not sufficient to prove Sternglass is right [but] the essential point is that Sternglass may be right. The margin of uncertainty in the effects of world-wide fallout is so large that we have no justification for dismissing Sternglass's numbers as fantastic."[12]

More recently the National Cancer Institute (NCI) released a thyroid cancer study in 1997 that estimated between 11,000 and 230,000 cases of thyroid cancer will likely result from radiation released during the atmospheric testing era. Many nuclear industry representatives and scientists dismiss the study's estimate as a vast overstatement, while many epidemiologists and downwind residents argue the NCI study significantly understates the scope of the thyroid cancer problem.[13]

Several impediments prevent making a concrete estimate of the scope of health problems related to Cold War radiation exposure in the United States. The first major obstacle is our incomplete knowledge of how much radiological pollution was actually released by weapons testing and the uranium industry. Technology for measuring the polluting by-products of nuclear tests was primitive at best in the early years of the Cold War, and the AEC tended to make such safety monitoring secondary to the military task of measuring the explosive effects of the weapons. Since scientists knew relatively little about the by-products resulting from the tests, the way they traveled through the environment, or their potential effects on humans, many toxic radionuclides were never monitored at all. Similarly, obtaining great quantities of uranium quickly and cheaply was more of a priority than was monitoring the contamination caused by the uranium industry.

The second major hurdle in assessing the scope of radiation-related illness is incomplete epidemiological data. *Epidemiology* is the science of examining patterns related to human health. One can discern patterns only with historical data about specific diseases among specific populations in specific areas, and the first nationwide system for tracking deaths was not established until 1949. Epidemiologists attempting to understand patterns of disease rely on accurate data regarding the cause of death, and many radiation-related ailments were not recognized as such by doctors and medical examiners in the early years of the Cold War. Even with all the data in the world, linking a specific person's disease to a specific causal agent is nearly impossible. A woman could smoke cigarettes for forty years and die of lung cancer, but scientists would still be hard-pressed to prove that woman's lung cancer resulted from her smoking habit. For decades, industry representatives have argued that if scientists or injured parties cannot prove that a product has caused harm in a specific case, then it must be considered safe, and this burden of proof remains the standard in American courts and public health policies.

While these impediments make assessing the true extent of radiation-related illness resulting from the Cold War difficult, they do not make it impossible. Radiation releases were monitored, albeit not in their entirety. Uranium toxicity at mines and mills was measured, if incompletely. Prior to the establishment of the national registry in 1949, death records were kept at the local level, providing data for some epidemiological comparison. More data was available from families, who often had ample information about the illnesses and potential exposure of their loved ones, if an epidemiologist had bothered to seek them out. Just as one can document the effect of cigarette smoke on the lungs and make an estimation of how many cigarette smokers might experience cancer, given a certain amount of exposure, one can document the effect of radiation on the body and thus estimate how many people are likely to develop health problems related to certain amounts of exposure to radiation.

From the inception of the nuclear weapons development program, U.S. government representatives were well aware that radiological exposure bore serious risks for citizens' health. Federal research agendas were established to document those risks as they affected the bodies of those people exposed to radiation, few of whom provided researchers with informed consent. American doctors and scientists began studying the health effects of nuclear weapons dropped on Hiroshima and Nagasaki almost immediately. By 1949 the AEC had established a domestic biomedical research agenda dedicated to monitoring the movement of radiation through the environment, the food chain, and the human population. Government scientists routinely measured radiation in both the milk supply and the thyroids of children in the immediate downwind region, collected tissue and bone samples from around the world, and in some cases experimented by directly dosing research subjects with radioactive substances.[14]

Early proponents of nuclear weapons testing were fond of pointing out publicly that radioactivity was an omnipresent natural phenomenon, emitted harmlessly and constantly from the sun and rocks. Humans had found many uses for radiation, these proponents noted, from luminous wristwatches to x-rays. Some even argued radiation exposure could offer regenerative health benefits and endlessly extolled the role of radiation in cancer treatment. A January 1946 article in the *Atlantic Monthly* reassured Americans that via "medical advances alone atomic energy has already saved more lives than were snuffed out at Hiroshima and Nagasaki." While radiation does exist in nature, radioactive pollution created by the nuclear arms race dwarfed any existing naturally or any received from wristwatches and x-rays, and it is possible that radiation has caused as many cases of cancer as it has treated. Some commentators compare the atomic bomb to Pandora's mythical box: once it had been opened, the danger it released could not be contained. Optimistic claims to the contrary, widely circulated in the late 1940s and 1950s, sought to allay the gloom and guilt that came with having invented (and deployed) the most dangerous weapon ever known to humankind.[15]

The lack of complete knowledge about radiation effects and the pressures of safeguarding national security do not absolve the federal government of culpability for endangering citizen health. In his 1984 opinion in *Irene Allen et al. v. United States*, the major downwinder lawsuit against the federal government, Judge Bruce Jenkins concluded: "The Government was bound by a legal duty to act with the highest degree of care in light of the best of available scientific knowledge." While recognizing that "knowledge and information was not constant; it was persistently expanding," Jenkins affirmed that "by 1950 . . . the concept that genetic alterations occurred at any dose of radiation went essentially undisputed among those concerned with radiological protection." In the early years of weapons testing, AEC scientists understood that even low doses of radiation could cause detrimental changes in the human body, yet the agency continued to base safety precautions on the concept of threshold doses—a flawed scientific assumption that radiation exposure is safe up to a point. Judge Jenkins concluded that AEC representatives acting on behalf of the government had "failed to adequately, contemporaneously, and thoroughly measure and monitor such fallout so as to be able to inform persons at risk of the extent of the hazard faced by each." The same point is unarguably so in the case of the uranium workers and their families, who were exposed in pursuit of the raw materials needed to construct nuclear weapons. The government knew serious health risks were involved, risks the AEC actively concealed from laborers even as a Public Health Service study conducted without laborers' informed consent raised warnings of imminent cancer.[16]

There is no reason to believe that living upwind from the Nevada Test Site or the uranium industry exempted one from the risks of radiation exposure. Uranium mines and mills and nuclear facilities dotted the entire country, from the Hanford Nuclear Reservation and Midnight Uranium Mine in eastern Washington to the Oak Ridge National Laboratory in Tennessee. Scientists have concluded fallout from domestic and international tests was eventually "distributed fairly uniformly over the entire United States." Barely fifteen years after the first nuclear explosion, "there was no place on Earth where the signature of atmospheric nuclear testing could not be found in soil, water, and even polar ice." Many of the radionuclides distributed during the Cold War still exist in our environment, where they will continue to burn through their radioactive half-lives for centuries to come.[17]

Even if it were possible to clean up all the radioactive pollution and nuclear waste on the planet, we would be unable to simply wait out the effects of this period in history. Claudia Peterson's daughter Bethany was born in 1981, nearly two full decades after nuclear testing went underground, ostensibly for the safety of the citizenry. In 1987 Bethany Peterson died of acute monoblastic leukemia. Her case might have seemed an isolated tragedy were it not for the cancers that also had plagued her mother, both of her grandfathers, and her aunt. Further, three other children Bethany's age were all diagnosed with cancer when she was, and they were all children of Claudia's classmates from East Elementary in Cedar City. Just as radioactive pollution insinuates itself into the atmosphere, water, and soil, so too does it insinuate itself into human genes, where it can cause lasting alterations. "Nuclear radiation is thousands of times more deadly than factory smoke," Professor of Environmental Politics Frank Fischer explains. "The cumulative impact of modern toxins on the human body and the wider ecosystem, largely unknown, is after a certain point of exposure irreversible. Moreover, such risks often respect no temporal boundaries. Across generations, they can accumulate in both intensity and complexity." Ultimately we all live downwind of a nuclear test, but some people live closer to ground zero than others do. For Claudia Peterson and every person who watched a mushroom cloud, thousands more have lived and worked in the path of that same cloud but have no stories about it because by the time it reached them, it looked like the other clouds in the sky.[18]

The downwinders and the families of uranium workers did not need conclusive science or historical documentation to confirm what their shared experiences led them to believe. While they had no way to identify or measure radioisotopes such as iodine-131 or strontium-90 as they breathed them in and consumed them in locally produced food, residents of the small communities east of the test site could see the mushroom clouds on the western horizon, and they became familiar with the cloud of dust that followed. The families of uranium workers were intimately familiar with the dust their husbands and fathers brought home from the mines and mills on their clothing, clothing that they wore as they held their children and that they gave to their wives to wash.

As patterns of health problems began to emerge in their communities, downwind residents and uranium families thought back to the specter of the mushroom cloud and to the stories elders told about the risks of mining into the earth. In the stories many began to narrate, the mushroom cloud changed from a symbol of national pride and strength to an icon of doom. Some Navajos began to refer to *leetso* as the "yellow monster." Everyone told his or her own version of the same story: good patriots all, the people had borne witness to the early years of testing or the uranium boom. No one had told them how dangerous it was, and the government that willingly exposed them had betrayed them in a way that they could scarcely express. Many storytellers began to use the phrase "guinea pigs" when discussing themselves and their neighbors. They described the rashes the dust left on their skin, the malformed lambs born to their sheep herds, the women who miscarried, and the men whose lungs gave out before they grew old enough to be grandfathers. The stories sounded like folktales, and the storytellers knew it. To convince their listeners of the veracity of their claims, narrators noted their histories in that place, pinned down events with dates and details of local geography, and shared their own theories on how local landscapes or weather shaped the path of fallout or uranium contamination. They reminded their listeners that everyone in their communities had relied on local agriculture, backyard vegetable gardens, local water, and fresh milk, all of which are vulnerable to radiological pollution.[19]

I first encountered these stories while doing research for a master's degree in history in 2004. I wanted to write about the experiences they described, but I was unnerved by the almost formulaic similarity between them. People employed the same plot trajectories, used the same phrasings, and tended to project the same emotions at the same points in their stories. Many conflated memories with details and facts they did not learn until decades after the incidents they described, making it challenging to separate memory, media revelation, and myth. Were these stories the elaborate conspiracy theories of people trying to make sense of painful losses, or were they historically accurate? When I began researching nuclear weapons development, I discovered that the "official" records of this era—those produced by government employees, scientific researchers, and courts—were riddled with as many agendas, inaccuracies, omissions, and rhetorical flourishes as the citizens' memories were. Every author seemed to use strategic fictions, and it began to seem that the truth lay somewhere in the middle. Clearly a profound contamination had occurred across the American West. While in some areas that contamination may not have been as dire as many had thought, in others it was likely far worse than anyone had ever imagined. The connection between that contamination and particular illnesses may be questionable in some cases, but undoubtedly the people who believe in that connection will never see their government the same way again. I concluded lingering questions of scientific proof and historical accuracy were less important than the shift in people's consciousness that had clearly taken place, fueled and documented by stories.

Folklore—which includes personal narratives as well as myths, fairy tales, jokes, and legends—is repeated and passed on because it serves a purpose. Stories do more than entertain; they explain particular customs or circumstances and communicate important messages about social norms. Many stories reveal for the audience a fundamental aspect of the narrator's history and identity, which often mirror in some way the history or identity of the audience. This feature is especially true of personal narratives. Derived from the Latin *narrare*, to "narrate" is simply to make known. Individuals share personal narratives, those dynamic compositions of memory and culture, to express vital truths about their life experiences. Like other forms of folklore, the personal narrative reflects a vital reality rarely explored fully in historical literature—namely, the experiences of ordinary people and the interpretations and values they ascribe to those experiences. Folklorist Richard Dorson calls these accounts "the chief available records for the beliefs and concerns and memories of large groups of obscured Americans." While these obscured Americans are often people living in poverty, people in rural areas, or people of color, more visible and ordinary Americans, such as white middle-class families living in cities, also note their experiences are overlooked in the history books. "We need these stories of communities," folklorist Sam Schrager writes, "because through them we can gradually come to inhabit a new story of this country, a story that will help us come to terms with our present circumstances and our unfinished past."[20]

Historians refer to personal narratives as oral histories, and their use of such histories as source material has become a well-established and respected methodology in recent decades. The printed documents of governments, courts, and academics are still privileged over the memories of ordinary people, who are broadly seen as being less reliable and authoritative. Historian Wulf Kansteiner opines that "while most academics still maintain that 'in its demand for proof, history stands in sharp opposition to memory,' there are good reasons to question such a clear epistemological divide between academic and non-academic representations of the past." While suggesting that "perhaps history should be . . . defined as a particular type of cultural memory," he also invokes the words of historian Peter Burke: "Neither memories nor histories seem objective any longer. In both cases we are learning to take account of conscious or unconscious selection, interpretation, and distortion. In both cases, this selection, interpretation, and distortion is socially conditioned." In other words, we cannot separate ourselves from the society and the era that enfold us at the moment we choose to consider history. We remember the details that seem important to us, and historians and people in power record the events that seem important to them. No one is capable of standing apart from this complexity to author the whole story of the past.[21]

Reflecting on four decades spent observing human society, anthropologist Clifford Geertz once wrote, "It is not history one is faced with, nor biography, but a confusion of histories, a swarm of biographies. There is order in it all of some sort, but it is the order of a squall or a street market: nothing metrical. It is necessary, then, to be satisfied with swirls, confluxions, and inconstant connections; clouds collecting, clouds dispersing." As much as students of history might prefer to tack those clouds onto a linear timeline, Geertz dismisses the possibility of securing an accurate, full recapitulation. "There is no general story to be told, no synoptic picture to be had," he writes. "What we can construct, if we keep notes and survive, are hindsight accounts of the connectedness of things that seem to have happened: pieced together patternings, after the fact."[22]

This book is precisely such a pieced-together patterning, which is made possible by the notes, the survival, and the hindsight accounts of the downwinders and uranium-affected people that are considered in tandem with the documentary record. Having realized that no single academic discipline would give these stories their full due, I approached them with the tools of history, folklore, cultural anthropology, environmental studies, and epidemiology. This book became a study in narrative method and historical documentation, memory and forgetting, loss and survival, the history of those in power and the history of those upon whose bodies and homelands that power was practiced. It is a study of that most fundamental of human inclinations—the attempt to make sense of our experience through story.

Long after the dust has settled and the politicians have claimed victory and the historians have penned their summaries, cancers and illnesses will continue to manifest in the bodies of ordinary people, erasing the supposed boundaries between soldier and civilian and making us all survivors and potential victims of a war we thought had ended. It is time for a new accounting of the Cold War, one in which the so-called folktales of the ordinary people who witnessed it firsthand are accorded the credibility and attention they deserve.

{2} UNEARTHING YELLOW MONSTERS

We would find these [yellow veins of mineral] like you found big dead snakes, because you could see where the head was and the tail was. . . . We would look at it and wonder what it was. We would say it must have been the ancient monster snakes from the beginning of time.

George Tutt, Shiprock, New Mexico

The place where a story begins relies entirely on the perspective of the storyteller. America's nuclear arms race is a story of many such perspectives. Some historians begin their narrative of the atomic bomb with a group of brilliant physicists puzzling over calculations in secretive laboratories. There are Utahans who will tell you that, for them, the story of the bomb begins on a day in 1953 with a brilliant flash in the morning dark, followed by a mushroom cloud rising in the western sky and slowly drifting toward their homes. For many North American indigenous peoples, the story of the bomb begins in the earth, and often it is not ostensibly about the bomb at all.

Thinking back to the World War II era, Tommy James of Cove, Arizona, reflected, "Back then in those years . . . jobs were scarce. That is how it was." The economic revitalization that the rest of the nation experienced during the war years and those that followed had not visited the Colorado Plateau's reservations, which were geographically and culturally remote from national industry. Native communities that called the region home had weathered generations of challenges, resisting and adapting to the violence and encroachment of Spanish, then American, colonization while continuing to develop their dry-farming and livestock-raising cultures in a region defined by extreme water scarcity.[1]

In order to thrive, these communities cultivated intimate familiarity with local landscapes, ecological conditions, and historic and seasonal climate patterns in the arid region they called home. A sophisticated folklife tradition emerged to preserve and transmit this familiarity to younger generations. Oral storytelling, handcrafts such as weaving and toolmaking, architectural styles, and agriculture and food-preservation techniques all integrated this experiential, local knowledge and enabled groups such as the Navajos and Puebloans (known today by the names of their communities: the Hopi, Laguna, Zuni, Acoma, Taos, and Jemez peoples) to persist and flourish.[2]

As the atomic era dawned, many Native people were struggling to keep this traditional knowledge alive in communities that had been fractured by the changes of the nineteenth and twentieth centuries. Several generations of Native children, many of whom had been forcibly removed from their families, by then had been educated in Bureau of Indian Affairs (BIA) boarding schools. Living away from their communities for months and years at a time, these children were prohibited from speaking their native languages or practicing their home customs, causing severe cultural dislocation even after they returned to their families. Farming, hunting and gathering, and livestock raising had been heavily curtailed by the federal imposition of reservation boundaries and livestock reductions, leaving many families unable to support themselves as previous generations had. Young people began leaving their home communities in search of wage labor, and many enlisted in the armed services during the Second World War.[3]

Some of those people who remained on the reservation found income working as guides and laborers for geologists and fortune hunters searching out uranium for the federal government. The demand for the element came from the Manhattan Project, a highly secretive task force convened in 1942 under the direction of Maj. Gen. Leslie Groves and physicist J. Robert Oppenheimer. Their goal—to build a bomb of unprecedented power, fueled by a nuclear reaction—hinged on their finding a particularly volatile natural element. The physicists surmised that if they bombarded that unstable element with neutrons, the atoms of that element would split into smaller particles, and in turn they would create fission, a chain reaction that the scientists hoped to harness to produce a nuclear explosion. Two isotopes of the element uranium—uranium-235 and uranium-238, which the physicists learned to synthesize into an element they dubbed plutonium—proved the ideal candidates for fissile reactions. At the outset of the Manhattan Project, most uranium was being mined outside the United States, and the ore in transit was vulnerable to theft and wartime disruptions. Locating domestic reserves of uranium became a high priority for those who believed atomic weaponry could win the war and ensure national security.[4]

Identified in southern Utah in the 1890s, the uranium deposits of the Four Corners region had received only minor attention in subsequent decades. Some demand existed for uranium oxide, a colorant used in porcelain and ceramic production, and in 1902, Pierre Curie and Marie Curie developed a method for isolating radium, subsequently used in cancer treatments, out of pitchblende uranium. These applications created commercial demand for a small amount of uranium, but the prospectors who flooded the Colorado Plateau region in the early twentieth century were looking for something else—vanadium, which is often found alongside uranium deposits. Eventually used as a strengthener in the production of industrial metals, vanadium was mined from the Colorado Plateau in significant enough quantities to merit the construction of the first vanadium-processing plant in the United States, sited in west-central Colorado in 1900. Many Native residents of the region had their first taste of wage labor while working as guides for vanadium prospectors and laborers in the mines. Frequently unearthed along with vanadium, uranium tended to be discarded. Timothy Benally Sr. recalled stories of "uranium in gunny sacks stacked in the trading post at Beclabito [just south of the Four Corners and east of the Carrizo Mountains]. It had been there for a number of years before they found out what it was."[5]

They found out soon enough. Once the Manhattan Project scientists deemed uranium the necessary element in their quest to create a chain reaction, government geologists and prospectors flooded the Four Corners region with their Geiger counters. They were participants in what historian Raye Ringholz described as the nation's "first federally-controlled, federally-promoted and federally-supported mineral rush." The Atomic Energy Commission set a guaranteed minimum price for uranium ore and offered tantalizing cash bonuses for individual discoveries. Additionally the AEC designated itself the only authorized purchaser of uranium, establishing a total monopoly over the industry and giving itself powerful leverage over the companies that sprang up to extract the sought-after mineral.[6]

Many Navajos had observed uranium while herding sheep or acting as guides for vanadium prospectors in years past. As noted previously, they called it *leetso*, which translates from Navajo as "yellow brown" or "yellow dirt." Directing the new prospectors to this yellow dirt was an easy way to profit, or so it seemed. Few Navajos spoke fluent English, and those who felt they were compensated unfairly for their services or discoveries had little recourse. Many traditionalists warned against aiding the prospectors, believing mining to be

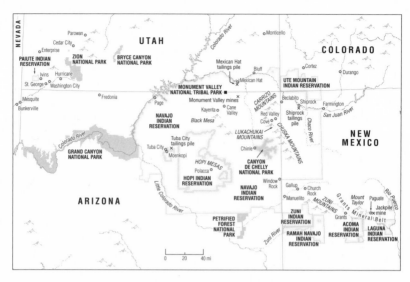

Map 2. Selected uranium sites and uranium-affected communities in the Four Corners region. Map by Erin Greb Cartography.

a transgression against the sanctity of the earth, but others found the allure of financial compensation difficult to resist. Despite the advice of his father, who warned him against assisting the white men swarming over their lands, sheepman Luke Yazzie took a sample of ore-bearing rock from near his home in Cane Valley, northern Arizona, to the local trading post, which was run in nearby Monument Valley by a white man named Harry Goulding.[7]

Within the year, Vanadium Corporation of America [VCA] had taken out a mining lease on the land near Yazzie's home and initiated a major uranium-mining operation. Years later, Luke Yazzie's nephew Ben Stanley recounted his grandfather's advice to his uncle: "He said, 'Never take [the uranium rocks] to the white man. If you do, you'll get nothing out of it.' Luke took it to Goulding. He got a cigar for it." Goulding described himself as a friend of the Navajo people, as well as an "interpreter, amateur doctor, general advisor, intermediary on legal matters, lodge operator, prospector, and a one-man chamber of commerce." He saw the development of mining in the region as a major financial opportunity and not only for the Navajos, whom he designated himself to represent. Goulding drew regular checks for "prospecting" from VCA for years, and the Navajo laborers spent much of their wages at his trading post.[8]

Four decades later, journalist Peter Eichstaedt interviewed Luke Yazzie about the changes his decision had precipitated. With his daughter and daughter-in-law translating, Yazzie described in Navajo how his family's quiet valley filled with eager workers who sought wage labor in the mine. The piñon and juniper trees were harvested for firewood, and before long, hundreds of people were camped nearby. The mining company hired interpreters to translate for their vast pool of Navajo laborers, and a store opened to sell goods to the workers. "The workers were told they would get a portion of the profits," Yazzie told Eichstaedt, who recalled that "with disgust in his voice, [Yazzie] said, 'while we were working, Mr. Goulding got rich and moved to Phoenix.'" Not long after Luke Yazzie brought the VCA to his backyard, Navajo sheepherder Paddy Martinez led geologists to a uranium vein that was quickly revealed to be a piece of a massive deposit dubbed the Grants Mineral Belt. Mining and milling activity exploded in the region.[9]

Aware of the economic opportunity connected to the mineral on their land and eager to demonstrate their patriotism in the burgeoning Cold War effort, the Navajo Nation Council passed a resolution in October 1949 supporting the development of the tribe's uranium resources. Tribal leaders and BIA officials tried to prioritize Navajo rights in uranium prospecting and mining, but these rights were routinely exploited. William Chenoweth, a geologist who worked in the region for the AEC in the 1950s and 1960s, described a common scenario to Peter Eichstaedt: "Here'd be a family herding sheep in this area, having their hogan in this area, and maybe a white prospector would find an outcrop up near their place. And he'd need a mining permit." According to terms established by the BIA, only

Navajos were eligible to apply for mining permits, but as Chenoweth explained, "you could use any Navajo in the world. So here'd be this family living here on this land for many, many years. But if you knew an English-speaking Navajo . . . you used him to get the permit. And if a mine was developed, your permittee, your Navajo partner, would get [a] royalty (2%), but the people living there wouldn't get a thing."[10]

The pattern Chenoweth described played out in numerous mining booms on reservations across the United States. In order to extract resources from reservation land, large mining companies such as Anaconda and Kerr-McGee first needed to negotiate access through the BIA, which holds reservation lands in trust for Native Americans. Because European American society has for many years considered Native peoples incapable of managing matters of property and financial negotiation, the BIA acts on behalf of the tribes it oversees in what is known as a fiduciary, or trust, relationship, which is the same legal arrangement used to protect and manage the assets of children. Though this fiduciary relationship at times has protected Native interests, historically the BIA has often catered to industries, greasing the wheels for resource extraction companies' profit making at the expense of those to whom the land legally belongs.[11]

Acoma scholar and longtime indigenous rights activist Manuel "Manny" Pino described the process of mining lease negotiation, which the BIA still manages in the twenty-first century: "You'll have three entities sitting at the table. You'll have the mining company, the tribe, and the federal government as our trustee." Noting that "the royalty payments, the lease payments, [are] being set at fixed rates, as opposed to taking into consideration the fluctuation of the market," Pino exclaimed incredulously, "in many of these cases you'd think that the federal government is trustee of the mining company rather than [of] the Indians." He mused that at the outset of the uranium boom, "Indians at the time didn't understand the capacity of this type of development, the dangers of radioactive exposure. They took the government's word for it that this was going to be economic development, that this was going to be jobs, this was going to improve the peoples' quality of life."[12]

At the outset of the first uranium boom few Americans, let alone residents of remote Indian reservations who lacked access to a national media culture growing saturated with references to radiation and the atom, understood the factual particulars of radiation exposure. Even those Native peoples accessing the national media would have found little to no reliable data on the dangers posed by uranium extraction.[13]

The leases the BIA negotiated amounted to an informal subsidy for resource extraction companies. Native communities received royalties amounting to only a fraction of the market value of the uranium. As late as 1984, uranium royalties for indigenous peoples averaged less than 4 percent, with 15 percent being standard off the reservation at that time. Uranium leases also permitted mining companies to pay Native miners substantially less than what they paid off-reservation employees for comparable work. Despite these massive inequities, the income from the uranium industry proved difficult for many tribal communities to resist, particularly since wages could be earned close to home. Pay stubs of Navajo miners from 1949 show an hourly wage ranging from $0.81 to $1.00 an hour. While this sum seems paltry in the twenty-first century, roughly half of American families in 1948 survived on less than $5,000 a year in income, and conditions on the reservations were substantially further below the poverty line. (Economic conditions have not improved on reservations much since then. In 2001 the average per capita income on the Navajo reservation was $7,269, with 42.16 percent of the reservation's residents unemployed.)[14]

On the Laguna reservation in west-central New Mexico, wage labor jobs were just as scarce as they were on the nearby Navajo reservation. Manny Pino estimates that unemployment in Laguna was around 70 percent prior to the arrival of the uranium industry. In 1953, the Anaconda Mining Company began operating the Jackpile Mine on

Laguna land, and at the mine's peak, more than eight hundred Laguna people found work there, dropping the unemployment rate to less than 20 percent. Laborers did not have to travel far from home as the mine's edge lay a scant two thousand feet from the village of Paguate, a traditional Pueblo village dating back to the seventeenth century.

For generations, Laguna families had supported themselves as Alveno Waconda's did, "by making a living from farming and raising livestock. My early childhood days were spent helping my father." After his father took a job in the Jackpile Mine, Waconda's family "never farmed on a large scale basis again. My father sold his livestock because he could no longer tend to his livestock and all the work involved at the mine." With the family's farming enterprise thus reduced, Waconda followed his father into the industry when he was old enough to take a job in the mine.[15]

Across the region, Native men who had left their homes in search of work and Native veterans who had finished their military service returned and took jobs in the uranium industry, often moving their families to mining camps so they could live together as they had before. George Tutt recalled the uranium boom: "It was good! Work was available close to home. We were blessed, we thought. Railroad jobs were available only far off like Denver . . . but for mining, one can just walk to it in the canyon. We thought we were very fortunate, but we were not told, 'later on, this will affect you in this way. . . . We called it 'good work.'" Tutt's qualifying statement, "we were not told, 'later on this will affect you in this way,'" foreshadows the realization he and other uranium laborers came to years later: uranium work bore unseen dangers that would reveal themselves over time. In the interim, the only warnings about the industry came from traditionalists who opposed all forms of mining. To counteract these voices, trader Harry Goulding developed a spiel for miners who "expressed moral qualms about the digging." Drawing on the Navajos' reverence for another yellow powder—corn pollen, one of the central components of many Navajo ceremonies—Goulding told miners,

"The *leetso* is the same color as the corn pollen. . . . You could put 'em side by side and you couldn't tell 'em apart; they are the same yellow color exactly. We need that yellow dirt because it is strong like the pollen, and it makes bullets and helps in the war so much." Goulding's suggestion that uranium mining could aid in protecting national security resonated with his Native audiences, many of whom either had served in the military themselves or had a family member who was a veteran.[16]

Timothy Benally came home to the Navajo reservation in 1958 after his stint in the armed forces. The only job he could find was in the uranium industry, so he took it. Relating his story in 2008, Benally began by holding up four fingers. "In Navajo we say four is a sacred number. And that was my first time, or the first job, with the uranium."[17]

The structuring of stories on numbers significant to the story-teller's society is common in folklore from every part of the world. In fairy tales, wishes tend to come in threes, brothers and sisters in sevens. In Diné Bekayah, as the Navajo call their homeland, things tend to come in fours. For Benally, it is no coincidence that he ended up working four jobs in the uranium industry. The four directions govern Navajo conceptions of space and spirituality, conceptions intimately tied to the four mountains surrounding Navajo country that in turn play pivotal roles in the Navajos' story and are considered sacred. Folklorist Barre Toelken explains that the number four is crucial to "patterns of balance and reciprocation . . . important to Navajo mental health," and these patterns are enacted in ceremonies, dances, handcrafts, and oral narratives.[18]

The uranium industry threw Benally's life, his community, and the local environment out of balance. The restoration of that balance—in terms of emotional and physical health, spirituality, and ecology, all of which are interconnected for the Navajos—is crucial to the healing of all involved. By setting up the story of his mining career as a four-part narrative, Benally brought a discordant experience into the

Fig. 1. Timothy Benally, shown examining a historic photo of the workers of the Navajo Uranium Division, Kerr-McGee Company. Photo by Theresa Trebon, Shiprock, New Mexico, 2008.

structure of the Navajo worldview, facilitating a restoration of balance and the progression of healing. The structure of his story also served a rhetorical function; having set the stage for a four-part story at the outset, Benally created a narrative arc that helped to draw in his audience.

"I caught a ride with [my brother-in-law], and we went up to Cove [in northeastern Arizona]. And that was my first job with the uranium," Benally recounted. "We followed uranium from there to Monticello [Utah], and then we followed others to Durango, Colorado, . . . and after that, I didn't work with uranium again, until 1957, I think." Explaining that he and his brother-in-law migrated from mine to mine, Benally revealed they followed the dominant pattern in the industry, with itinerant laborers moving from job site to job site. Identifying Cove, Monticello, and Durango, Benally indicated he had worked at three of the most productive uranium sites in the Four Corners region.

Benally's second uranium job lasted "four or five months. The workers were complaining about the work condition. And I said, ... 'There's a way to solve the situation.' And they said, 'How's that?' And I said, 'We can unionize the workers, and when we did it, we'd have the bosses listen to us.'" That Saturday, the end of the work-week, Benally was given his pay and told not to come back to work. "I didn't think anything about it," he recounted later.

Benally's emphasis that he "didn't think anything" about losing his job was indicative of an attitude toward wage labor that is common among many itinerant workers in the uranium industry. The mining industry is, by its very nature, one that booms and busts as demand for the mineral changes, thus conditioning workers to see such jobs as temporary. Cultural factors made this attitude especially preva-lent among Native laborers. Toelken explains, "The Navajo concept of time is . . . not seen as a pathway along which one passes but . . . [as] a context in which things move about." He clarifies this concept using the example of a Navajo craftswoman: "Instead of planning the future [of the object she is creating] as if she were separable from nature, she negotiates with those forces presently around her; instead of looking toward a future product, she looks to the past for pattern-ing, for advice and wisdom." We can deduce that in this worldview, such concepts as long-term job security and saving money for the future would be illogical.[19]

Social factors underscored this reality. With few opportunities for steady wage labor in their home communities, many workers accepted the need to move from job to job, and few expressed the loyalty to any one company or job that was so prized in European American society. A 1961 report by Wilbert Dare for the Kerr-McGee Company indicated as much. "When Kerr-McGee first began mining on the reservation, the turnover of Navajo labor was very high, mainly because many of the men did not easily accept routine. Normal to their way of life, the Indians worked only long enough to satisfy their immediate needs," Dare opined. "Although the Navajo are conscientious and hard-working miners, it is only within the last few years that they have appreciated the need for reporting daily to the job." While acknowledging a solid work ethic among Navajo laborers, Dare fell back on a popular stereotype of Native men—"many of the men did not easily accept routine"—to explain the high turnover rates. On the contrary, Navajo men, as with most indigenous peoples, were quite familiar with routine. Raising livestock, cultivating crops, observing spiritual ceremonies and cultural cycles—all demanded strong accountability to routine but just not to the nine-to-five, Monday-through-Friday routine to which Dare and other whites were attuned.[20]

A year after Timothy Benally was let go from the mine, a friend told him that "there's work available, they need more people." "They don't want me! They already fired me," Benally replied. "No, this is a new company, Kerr-McGee," his friend told him. Benally's third uranium job lasted approximately six months. During that time, he recalled, "somebody brought in a paper that said they were talking about legal wage, that the government had set, at one dollar twenty-five cents an hour, should be the minimum that they should pay." Benally told his coworkers: "Hey guys, we're being cheated. . . . Kerr-McGee is paying us way less than what they're supposed to be paying us." Once again, Benally received what he was told was his final check at the end of that week. "I was told, 'You're being a rift.' And later I learned that Kerr-McGee and the tribe had agreed to not firing Navajo workers. So what they'd devised is this thing about rifting; so they won't get tied up with troublemakers, they use that. Anyway, they told me that, and again, I said okay." As Benally explained, Kerr-McGee's lease to mine on the Navajo reservation prohibited the company from firing Navajo laborers outright, but as was true in industries across the country, workers who advocated for fair pay or unionization were poorly tolerated. To get around the agreement with the tribe, Kerr-McGee stipulated that workers who caused trouble, or "rifts" between the workers and the company, would be let go.[21]

By the conclusion of his third stint in the uranium industry, Benally had become quite accustomed to the way mining companies treated their Native laborers. He described how he sought other work after Kerr-McGee had let him go but found "there was no other job besides the mining and coal. And here, I was a veteran, United States Marine Corps veteran." By mentioning his military service, Benally alluded to the fact that even the most honored members of his community had difficulty finding work in the late 1950s, underscoring the bleak outlook for wage labor in the region. As noted, veterans are venerated in Native American communities and are particularly cherished among the Navajo people. Many Native workers in the first uranium boom were World War II veterans, and many in the second boom, which began in the 1970s, were veterans of the Korean and Vietnam Wars. They frequently mention their military service in recounting their uranium stories, a biographical detail that not only underscores their patriotism but also amplifies their audience's impression that the government had betrayed some of its most loyal citizens. Their government had failed to warn them of and protect them from the hidden dangers of the uranium industry even as the fruits of their labor were being used to produce weapons for the Cold War. This injustice is subtly addressed in a red, white, and blue bumper sticker seen on the Navajo reservation that reads:

Cold War Patriot
Uranium Miners
We did our part to keep America free!

In 1960 Benally took his fourth and final job in the uranium industry, toiling in a mine in Colorado. He worked there with his uncle for roughly seven months. At the time of our conversation in 2008 Benally told me his uncle was "dying from the effect of the uranium right now. He's just laying in bed, and he's on oxygen, that's how he can breathe. He and I were working for a man named Adolphus Moore. He was doing scavenging work, so scavenging leftover uranium that had been dug up." Benally described how in the early years of the uranium boom, "when the big companies went in, they just took the core of the deposits, and they left even heavy uranium as pillars in the mines." Adolphus Moore "wanted to take those out . . . he lowered a bucket down into this hole in the ground and [Benally and his uncle filled] it with ore." When they had filled the bucket, the men would ring a bell, and Moore would hoist the bucket out of the hole. "That's where I had an accident," Benally recounted, setting the stage for the final scene. He described how as the bucket full of ore was being hoisted up, a piece of ore "fell off and set right in the middle of where the bucket was supposed to sit. I guess it wouldn't have mattered if we didn't pick it up, but my uncle said, 'Jump in there and throw it out.' So I jump in as soon as the bucket disappeared. I thought I had plenty of time. Just as I got down there I see this dark vision coming down. I look up, and here comes the bucket. The next thing I saw was stars." Although he was not severely injured, Benally decided he was done with the industry. "And I said, this is it. I'm not gonna work in uranium mines anymore. We just said good-bye to our employer, and he hired some other people that were in need of a job. So that was the end of my uranium career."

Benally's story to this point could be the story of a laborer in nearly any blue-collar industry: poor pay, bosses intolerant of labor organizing, and dangerous working conditions. As Benally and other laborers eventually discovered, there was more to uranium than met the eye. Decades prior to the Cold War uranium boom, studies of European uranium miners had left "no scientific doubt that uranium mining was associated with high rates of lung cancer. The debate turned to identifying the causal agent," writes Doug Brugge, a professor of public health. Cognizant of these risks, the U.S. Public Health Service (PHS) embarked on a study of "radon in mines and health outcomes" in July 1950, "cooperat[ing] very closely with the Atomic Energy Commission."[22]

Worried that alerting the laborers to the risks of the industry—directly or via the recommendation that protective equipment be worn—would result in a diminished workforce, the mining companies, the AEC, and the PHS kept laborers in the dark even as they conducted physical examinations of affected workers, monitored environmental conditions in mines and mills, and documented emerging ill health effects. This maltreatment was not the case exclusively for indigenous miners; the industry and government agencies failed to share the health risks with white and Hispanic miners as well. The decision of AEC and PHS authorities to proceed in their investigations without informing their research subjects of the risks directly violated the 1947 Nuremberg Code, which mandated informed consent be obtained from persons involved in medical research.

"We did not want to rock the boat," PHS official Victor Archer later stated. "We had to take the position that we were neutral scientists trying to find out what the facts were, that we were not going to make any public announcements until the results of our scientific study were completed." Sociologist Susan Dawson suggests that in order to obtain the names of potential research subjects, the PHS made an agreement with mining companies "not to divulge the potential health hazards to the workers while they were monitoring their health, nor to inform those who became ill that their illnesses were radiation-related." Scientists responsible for public health thereby knowingly sacrificed the health of their subjects—and many of their subjects' wives and children—in order to obtain research data.[23]

As early as January 1951, an internal meeting of the AEC and PHS regarding Colorado Plateau uranium mines "reveal[ed] that the staff believed, based on early measurements, that radon was present in levels that would cause cancer, and that ventilation could abate the hazard." Brugge and Rob Goble note that "public acknowledgment of this problem was apparently squelched." The following year, senior sanitary engineers Duncan Holaday and Henry Doyle and surgeon Wilfred David wrote "An Interim Report of a Health Study of the Uranium Mines and Mills" for the PHS. The report did advocate the provision of rudimentary protections for the laborers. "Although the study is far from complete," the authors indicated, "sufficient information has been derived to conclude that certain acute conditions are present in the industry which, if not rectified, may seriously affect the health of the worker."[24]

Rather than warning the workers directly, the PHS recommended safety measures to the AEC and the mining companies, neither of whom had a vested interest in worker safety. For the AEC, obtaining uranium for weapons development was a matter of national security, and for the mining companies, obtaining uranium was a matter of profit. The first time workers appear to have been alerted to possible danger was in 1959, when "there were some pamphlets that mentioned a risk of lung cancer given to miners." Brugge and Goble argue these pamphlets "minimized the level of concern" and point out "it is unclear how widely these materials were disseminated, or what was the literacy/English comprehension of the miners who received them." In failing to inform the workers of the risks to their health in a timely manner—and in a language they understood—the AEC, the PHS, and the mining companies sacrificed the health of the workers' families along with that of the workers themselves. "When you purchase a prescription, there is a warning on the drugs. Warning to keep it out of reach of children, and to hide it—they make this clear on the prescription," Lorraine Jack, wife of uranium miner Leroy Jack, reasoned in 1995. "What if in the beginning [the employers and government] . . . had not allowed the children to come near the mine sites? The men were sacrificed already, it seems; they already knew, perhaps—they just watched us expose ourselves. . . . They should have told us, 'No.'"[25]

It is chilling to read the Public Health Service's 1952 report, which details in certain terms the process by which workers—and their families—were being exposed. Radon is a gas emitted by the decay of radium, a radioactive element found in uranium ore. As time passes, radon decays further into isotopes called radon daughters; these decay further into radioactive alpha particles. "Under usual mine conditions," the report explained, "large numbers of dust particles and water droplets are present in the atmosphere to which the solid decay products of radon will become attached. This dust will be inhaled and carried into the lungs where a portion of it will be retained and decay . . . thus delivering additional radiation to the lungs." The PHS report described alpha particles as "very energetic" and able to "damage body cells with which they interact." Despite their potency, alpha particles "can be stopped easily by such things as a sheet of paper or even human skin," Eichstaedt notes. Cheap precautions such as face masks could have offered miners significant protection. Absent such protection, "once alpha particles are taken into the human body, they lodge in tissues, bones, or organs, and steadily radiate and pelt surrounding cells."[26]

The dust produced by the mining process posed its own risk to miners' health, independent of any radioactivity. Most of the ore-bearing rock was sandstone, of which silica is a major component. Pulverized in the extraction process, it created great quantities of fine dust. The PHS study considered the risk that miners might develop the lung disorder silicosis from their dust exposure but concluded that "the silicosis problem in the mines is not acute" and "[d]ust control in the mines has been fairly good."[27]

Miners recalled a very different reality. George Tutt related, "There was no protection of any kind. You know, like a mask or anything. . . . They just blow, as they drill, they just blow the dry rock right in their face. . . . They were just covered with dust." Alveno Waconda also remembered "poor ventilation" and "dusty conditions." In 1992 he testified that "in the mornings, after waking up from a shift, I always remember coughing up the phlegm and seeing it black as the result of all I breathed in the night before."

In the case of this discrepancy, should one rely on the "expert" assessment of the Public Health Service or on the oral narratives of the miners? Time, and the ailments in their bodies, revealed the miners' accounts to be the more accurate version. Douglas Zang, a medical doctor working on the Navajo reservation, screens veterans of the uranium industry for health problems. "Most of what we're finding now is lung disease," he reflected in a 2008 interview. "We actually don't find [lung cancer] anymore . . . because they've all gotten it and died. What we find now is a form of silicosis, because the uranium deposits in this area are blasted from sandstone. . . . Without a mask, [miners were] essentially inhaling powdered ground glass." This material caused progressive scarring in their lungs, gradually diminishing the individuals' ability to breathe. Eventually, the depleted capacity of the silicosis patients' respiratory systems left them unable to fend off other ailments. Most of Zang's silicosis patients succumbed to "pneumonia, influenza, [or] congestive heart failure."[28]

Other toxic substances were inhaled besides the dust and the radon. Tommy James recalled, "There was one thing I reacted to, and that was the powder smoke [from blasting rock to get at ore deposits]. It was really bad. If the smoke was still in the air and you walked into it, it gave you a big headache." Blasting belowground caused a buildup of this smoke, abated to some extent by a rudimentary ventilation system in the mines. James remembered occasions when the ventilation fan stopped and the smoke filled the mine tunnels, a condition he claimed often resulted in nosebleeds. "Some of the mines were deep, and some of the men who were working in the tunnel ended up with no air. It was said the men were falling over."

James worked initially for the Sidney and Delaney Company, which subsequently sold to Kerr-McGee. "These companies were really alike in the work. It was like slave work," James told interviewers. Miner Logan Pete had similar memories of Kerr-McGee. When asked if he was told to wear face protection, he stated, "None, they did not tell us to wear such things. When the explosion of dynamite did not take place [fully], whatever was in it that did not burn would be so smelly. It got us all very sick for some time." In direct contradiction of the AEC's claims in the PHS report, Pete recalled being ordered to "hurry up and enter the mine [after blasts]. 'If you do not dig out all the stuff, you will be standing there digging again tomorrow,' we were told. Kerr-McGee treated us bad almost as though we were slaves."[29]

As the report makes clear, Public Health Service researchers did not ask the miners what went on in the mines; instead they asked the AEC. "According to the Atomic Energy Commission, generally only one shift is employed, and the mines are not worked on a round-the-clock basis. Consequently, workers are not exposed immediately following the blasting, when dust and radon concentrations are generally highest." Again, miners' narratives tell a different story. In her study of more than fifty uranium workers, Susan Dawson found numerous miners who recalled being "forced to enter the mines directly after blasting, when the mine was filled with smoke and dust. White workers were not forced to do so, according to the respondents." Even if the mines had blasted only when no laborers were present, blasting still affected those living downwind of the mines. In 1992 Alveno Waconda testified that "nearly every day" the Anaconda Company operating Jackpile Mine "would blast two times a day or more to get to the ore. There were times when the dust blew in our direction at work and many times toward the village of Paguate. Dust was in the air, settled on the soil, crops, clothes and home."[30]

"Radioactive dust is also concentrated on the workers' clothes and thus increases the workers' total apparent radiation dose," noted the authors of the PHS report. "The problem of measuring external radiation requires considerable study, but this inquiry has been deferred until other more acute problems have been solved." In the meantime, the PHS advised, "it is strongly recommended that the workers take daily baths (preferably showers) using soap freely, and that a frequent change of work clothing is desirable to minimize skin contact with the radioactive dust." The daily reality of indigenous uranium laborers afforded no such luxuries even decades after the PHS report. Tommy James recalled, "For the Anglos who were surveyors, mechanics, and the ones who worked in the office, they had showers for them. They had the mess hall for them. . . . For Navajo miners there was nothing, no facilities."[31]

This lack of facilities extended to even the most basic of workers' needs—clean drinking water. Having gone into the mines in 1955 after his military service, George Lapahie related, "We would get tired. We would get thirsty. We would drink the water, which was flowing down off the rocks in the mine. That was how we worked." Another miner, Phil Harrison, recalled that he "worked almost five months in a mine with my father, and almost every day I drank that water." Harrison described how the workers constructed a homemade cistern in the mine, with small cups resting on its edge. "So when you're working, and I used to help the drillers, bringing in dynamite, fertilizer, [you would] go over there, get like three cups, go back to work. Like going to a water fountain."[32]

Given that no clean water was supplied to the miners, it is not surprising that they were unable to obtain the "frequent changes of work clothing" on-site that the PHS report suggested to its small audience. Nearly every one of the uranium workers Susan Dawson interviewed "wore their dirty workclothes home for their spouses to wash with the family laundry. Some of the spouses who washed the family clothing said that they had rashes after finishing the laundry." Navajo miner Earl Saltwater Jr. told an interviewer: "These miners, they don't wear pajamas. Their working clothes were their pajamas. They touched their women like that. I hate to say it, but it's true." Lorraine Jack argued, "They should have provided washing machines so the miners could wash their clothes. Instead it was like herding sheep into a field of stickers. Right from the beginning, they should have told the families to live away from the mine site. . . . Children were exposed to it, and it contaminated our food and our drinking water." In her study "Navajo Uranium Workers and the Effects of

Occupational Illness," Dawson confirmed Lorraine Jack's memories, writing that many families lived in improvised camps in the immediate vicinity of the mine and "used water from the area for drinking, bathing, washing, and household uses."[33]

Those miners whose families didn't camp at the mine still exposed them to mine contaminants and not just through their dirty clothing. Phil Harrison described how "on weekends they used to pour [mine] water in canvas bags and take it home. Some of that water was used in the coffee." He remembered hearing "many stories about the babies dying. We lost my little brother when he was six months old. Using mine water in baby formula. . . . They were mixing it and feeding the babies. And my brother died, six months old, he died from a stomach ailment. So my mother, I guess she actually killed him with the mine water." The tragedy Harrison described has played out in many communities where formula has replaced breastfeeding as the primary source of infant nutrition. Without access to clean water, formula becomes a delivery mechanism for the toxins borne by contaminated local water supplies, a phenomenon that has been documented from Cove to Boston to Bangladesh.[34]

Mine contaminants found other ways into local water supplies. Donald Yellowhorse of Cove observed how as "waste and ore . . . were hauled out, some of these [were] washed down with water. . . . Then they hauled it over the roads, uranium fell off the trucks, so this makes uranium everywhere." Yellowhorse continued, "Dig wells, you drank the same water. They just dumped the waste into the washes, they flow down, that way everything was exposed, animals as well. This is what I think. Even down to the babies are like that."[35]

Decades later, researchers continue to confirm persistent radioactive contamination of local water supplies in the American Southwest. A 1999 report found that despite clean-up efforts, "ground water contamination continues at most" of the uranium sites studied, "including three of the four sites on the Navajo Reservation and the Falls City, Texas site. Active ground water remediation continues at

Monument Valley, Shiprock and Tuba City on the Navajo Reservation where contamination—including uranium, selenium, radium, cadmium, sulfate and nitrate—continues to exceed clean-up standards." The 2007 results of the Church Rock Uranium Monitoring Project found "*none* of the 17 sources tested were of sufficient quality to warrant recommendations for human drinking water use. Half of the water sources tested were *not* recommended for domestic uses, such as cooking, bathing and irrigating gardens, and most of the other half were recommended only with caution" (emphasis in original).[36]

The contamination that occurred in the course of removing the uranium from the earth was only the beginning. Subsequent to mining, laborers loaded the uranium ore onto trucks—frequently without covers—for transport to milling facilities where it would be refined. Manny Pino's grandmother lived within a quarter of a mile from the route that uncovered trucks and railroad cars fully loaded with uranium ore traveled "every day for thirty years," leaving behind a scattering of radioactive dust. Pino's grandmother succumbed to stomach cancer at the age of seventy-eight, becoming the first documented cancer death on his father's side of the family. The truck drivers transporting the ore were later found to be prone to elevated rates of renal cancer; chronic renal disease, including nephritis and kidney tubal tissue injury; lung cancer; and other respiratory diseases.[37]

Once the ore reached the mills it was pulverized, creating vast quantities of fine, radioactive silica dust, which was similar to that encountered during blasting in the mines. Subsequent physical and chemical processes were used to "strip, dry, wash, remove impurities, and roast the product result[ing] in what is known as yellowcake." The 1952 Public Health Service report noted, "Dust control at the crushing operation has been found to range from fair to poor. . . . Until adequate dust control has been established . . . workers should be required to wear approved dust respirators. Daily baths and frequent changes of clothing by the workers in this area are also indicated." In a survey of uranium workers' workplace conditions, millers informed

Susan Dawson that "no personal protective equipment" had not been provided "prior to 1970." As was the case with the mines, mills did not provide indigenous workers with bathing facilities on-site, and many millers lived with their families in rudimentary shelters in temporary camps, where the sort of thorough bathing that the PHS recommended was impossible.[38]

After a stint laboring in uranium mines, Tommy James became one of the first men hired to work at the milling facility in Shiprock, New Mexico, one of the largest communities on the Navajo reservation. He helped load ground ore into a tank, where it was mixed with sulfuric acid and water and heated to facilitate the extraction of uranium. James emphasized the company did not provide workers with protective equipment: "No face shield or anything like protective glasses." The PHS investigation into uranium milling found "relatively high concentrations of uranium and vanadium fumes . . . around the fusion furnaces. . . . In practically all plants the workers in this area were found to be suffering from a chronic irritation of the upper respiratory tract, apparently resulting from exposure to vanadium fume[s]." Other dangers plagued workers, and James recalled one instance when the sulfuric acid mixture "splashed out and I burned my face. They were small splash-spot burns, and it turned into pus, leaving scars after it healed. This is what the employees suffered from." James also described how the acid wore away the rubber coating on the propellers used to agitate the mixture in the tanks: "When the rubber coat wears off, then the metal becomes bare. When this happens, the acid eats the metal." James and his coworkers were tasked with repairing the propeller inside the agitator tank. "We had to sleep there in the tank when we worked on it, because it took so long to repair the propeller. And when the mill started back up [after the tank was repaired] the waste was spilled out and it formed a pond next to the mill and the waste all went into the river."[39]

The extensive refining of uranium ore produced a relatively tiny amount of usable uranium, known as yellowcake, but a massive quantity of leftover material, known as tailings. Scholar Dorothy Nelkin notes that only four pounds of yellowcake can be extracted from one ton of ore, and the leftover tailings "retain eighty-five percent of the radioactivity of the original ore." These tailings, deposited in settling ponds, usually remained near the mill. From these ponds, Arjun Makhijani and Scott Saleska note, "the water content . . . evaporate[d] or percolate[d] into the soil; the solid portion of tailings continue[ed] to accumulate." These tailings constitute "over ninety-five percent of the total volume of radioactive waste from all stages of the nuclear fuel cycle. If not isolated, they will release radioactivity into their surroundings for hundreds of thousands of years." Accumulated alongside the mills, these tailings piles became small mountains of uncovered radioactive refuse that gradually scattered throughout local landscapes and communities via wind and water runoff. Slowly this waste was absorbed into the food chain, a fact confirmed by Katherine Yih and her fellow researchers, who documented high radionuclide levels in livestock, plants, and wild animals near mills and tailings piles.[40]

Terry Yazzie related stories of his childhood living next to a massive tailings pile in Cane Valley to journalist Peter Eichstaedt. Yazzie said, "We used to play in it . . . we would dig holes in it and bury ourselves in it." Violet, his wife, remembered how Terry took her to visit the pile on her first trip to meet his family. "The kids played up there in the summer," she recalled. "We sledded on that during the winter. . . . Nobody said nothing about it from the time they stopped working [in the late 1960s] until the cleanup." The Cane Valley mine and mill that produced the tailings piles was established in 1955, after Terry's father, Luke Yazzie, led Vanadium Corporation of America geologists to a uranium vein near his home. "One of the biggest-producing and richest uranium [operations] on the Navajo reservation," the mine and mill shut down in 1967, forcing area residents to seek other employment.[41]

The tailings piles left behind in Cane Valley became part of the landscape, and local residents discovered the material was the perfect texture and consistency for use in the construction of their homes. Generations of Native residents of the Four Corners region have constructed their homes by hand, using mud mixed from local dirt as their primary building material as there are too few trees to provide lumber. The mud plaster serves to keep homes cool in the summer and warm in the winter. *Los Angeles Times* columnist Judy Pasternak relates how in 1975 while inspecting abandoned uranium worksites in Cane Valley, Environmental Protection Agency (EPA) radiation expert Joseph Hans, Jr., noticed "a small community of hand-built houses nearby . . . [and] returned with hand-held radiation scanners, air samplers, and other equipment." Hans documented that nearly half of the thirty-seven homes had been constructed with "tailings and/or uranium ore. . . . Tailings were used in concrete floors, exterior stucco, mortar for stone footings, cement floor patchings, and inside as cement 'plaster.'" Radiation was found in foundations, in walls, and even in a fireplace, as well as in other buildings not used as primary dwellings. Hans's measurements indicated residents of the dwellings had received dosages as high as 332 to 402 milliroentgens per year, or "triple the annual exposure of the average American." Locals living in proximity to the remaining tailings had likely inhaled "a high level of radon gas" from the decaying uranium piles. Two years later, the Utah engineering firm Ford, Bacon, Davis conducted a survey for the U.S. Energy Research and Development Administration and reported: "The presence of the tailings can be expected to increase the rate of occurrence of lung cancer by eighteen percent for those persons presently within one mile of the pile." However, the survey concluded, "as a result of the low population density . . . the potential health impact of the piles is minimal." The argument that radiological impact on human health is "minimal" because an exposed area is sparsely populated surfaces repeatedly in the saga of American nuclear development. Low population density is almost always a contributing factor in selecting sites for nuclear testing, nuclear waste storage, and uranium processing. When the people who do live in the area are not warned of the dangers, they are essentially being written off as collateral damage in the event that contamination occurs. While a few cases of lung cancer in a particular area may not be significant to a statistician, it is impossible to find a family who considers the impact of even one case of radiation-related cancer as "minimal."[42]

Mary and Billy Boy Holiday of Oljato, Utah, slept for years on a radioactive cement floor. A friend had covered the dirt floor of their small home cheaply by making concrete out of the sand and rock fragments left near an old uranium mine on the mesa near their home. Pasternak, who reported on the plight of the Holiday family and others, wrote in 2006: "By the time of the discovery that now torments her, [Mary Holiday] had lost her husband, Billy Boy, to lung cancer and congestive heart failure. He didn't smoke, but he'd worked in uranium mines by day and slept, unknowing, in the equivalent by night." Unaware of the danger for over three decades, his wife and three children "and a host of other kin" had slept alongside him. Similar problems have been noted on the Spokane Indian Reservation in eastern Washington, site of the Midnight Uranium Mine, which operated from 1955 to 1981. Unaware of the lingering radioactivity in the gravel and rocks left behind at the mine, locals returned to the area to haul away loads for driveways, foundations, and other construction projects.[43]

Local residents were only one of the forces carrying tailings away from the piles. Wind, rain, and floods moved the tailings into the local water supply. Evaporating from the surface of ponds and rivers, the radiation-bearing water droplets were carried into the atmosphere and returned to the earth as rain and subsequently were reabsorbed into the water table. This cycle cannot be interrupted, and it continues to the present day. The EPA has found contamination of groundwater at "virtually all U.S. mill sites," contamination that Yih and others point out "may be permanent." This contaminated groundwater then migrates from area to area. A toxic plume emanating from the groundwater near a tailings pile in Tuba City, Arizona, "has been reported to be moving 8.5 meters per year toward the drinking water sources of Tuba City and Moenkopi. The Moenkopi Wash is the Hopi people's only source of water for irrigation. The contaminant plume from the tailings pile at Church Rock, [New Mexico,] with elevated levels of arsenic and thorium, is believed to have reached the Navajo Reservation." Spatial borders such as reservation, state, and international boundaries are irrelevant to air and water, implicating human populations and ecological communities a great distance from the actual polluting activity.[44]

At many mills, tailings ended up in unlined holding ponds, part of a toxic sludge of water and chemicals that resulted from the milling process. Liquids evaporated from the surface of the ponds and leached into the water table beneath. Periodic flooding and structural weakness caused breaks in the dams holding back the settling ponds, releasing the toxic contents into nearby watersheds. The most significant break occurred on 16 July 1979 at a United Nuclear Corporation tailings pond at Church Rock Mill outside of Gallup, New Mexico, in the immediate vicinity of the Navajo, Zuni, and Acoma reservations. The accident released "1100 tons of radioactive mill waste and 95 million gallons of mine process effluent" into the tiny Puerco River, which then fed into the Rio Grande. "Seventeen-hundred people, mostly Navajo, live in the Puerco valley and about 2,600 head of livestock drink from the river," according to Nelkin. Resident Rita Begay remembered the river "had a terrible odor and a dark chocolatey color. Right away, we could tell it was unusual. It was a day later that it came over the radio about the spill, telling people to stay away from the river. A week later they put up those signs saying it was dangerous." The United Nuclear Corporation spill released 46 curies of radiation but received virtually no press coverage. The notorious accident at Three Mile Island Nuclear Power Plant, which occurred only a few months prior, released 13 curies. The disparity in reporting on these incidents is no doubt due to the fact that the tailings spill occurred in a rural, low-income area and affected a primarily nonwhite population.[45]

In the arid lands of the Colorado Plateau, waterborne contamination spreads swiftly through the food chain. Without an alternative water source, livestock and humans who relied on the river received significant doses of radiation as a result of the United Nuclear Corporation accident. Manny Pino explained that many area residents sold meat from their livestock at "meat markets in Gallup, and after the spill all the meat buyers would say, 'Where you from?' And if they said, 'Church Rock,' or if they said, 'Manuelito,' [the buyers would say,] 'Forget it, we're not buying your meat.'" Despite its ranking as one of "the largest releases of radioactivity in the history of the nuclear industry," Paul Robinson notes that the United Nuclear Corporation spill "does not appear to have had as devastating an effect on the Rio Puerco as the decades of mine dewatering which preceded the spill."[46]

In the scattered communities of the arid Four Corners region, water and wage labor have historically been equally rare, making both prevalent themes in the observations, memories, and stories of local residents. As laborers and the families of laborers, they became intimately familiar with the work of extracting and processing uranium. As residents of an arid region, by necessity they knew the locations and patterns of local water, an awareness that attuned them to the intrusion of mine and mill wastes into their water supplies. And despite their declining involvement in livestock raising and agriculture brought about by their increased participation in the wage labor market and by federal policies, these individuals were still members of agricultural communities, where the links in the food chain were commonly understood. While they could not discern the radiological contamination of their food chain, they could see the dust that floated out of the mines after blasts, that blew off of uncovered trucks transporting ore to the mills, that settled on their crops, and that washed away in the creeks where they watered their livestock. They knew it was the same dust that miners inhaled in the mines and coughed up the next morning, the same dust that came home on miners' clothing and left rashes on the skin of the women who laundered it. In the coming years, as strange and serious new health problems began to plague the uranium workers, their families, and their livestock, many began to question the wisdom of unearthing *leetso*.

{3} HOME ON THE RANGE

[W]e knew what happened because running sheep your whole life, it was something that never happened before and it's never happened since.
Annie Corry, Cedar City, Utah[1]

In the months and years that followed the bombing of Japan, accounts proliferated of the incredible potential of nuclear technology to both create and destroy. National and local newspapers and magazines reprinted verbatim the press releases, opinions, and fictionalized future war scenarios of politicians, scientists, and military leaders, stocking the American lexicon with a set of images and terms heretofore reserved for science fiction. Apocalypse, in the shape of a mushroom cloud, was understood to be on the doorstep, alongside incomparable power and possibility.

William Laurence, the *New York Times'* first ever science reporter, educated millions of Americans on the workings of the bomb. He did so in two capacities — that of a science reporter and that of a secret employee of the U.S. government. Laurence wrote about the power and promise of theoretical nuclear technology through much of the 1930s and 1940s, and in 1945 General Groves of the Manhattan Project approached him about becoming its publicist. In an arrangement kept secret for decades, Laurence purported to offer objective, science-based journalism about nuclear developments, all the while feeding the public Atomic Energy Commission propaganda. Laurence's celebratory and awestruck descriptions of the Trinity nuclear test and the bombing of Nagasaki (which he witnessed from the air above) certainly cast the nation's nuclear endeavors in a positive light. "One felt as though he had been privileged to witness the Birth of the World—to be present at the moment of Creation when the Lord said: 'Let There Be Light.'" His articles and the stories related by the men aboard the *Enola Gay* who bombed Hiroshima established, in historian Patrick Sharp's words, "the narrative sequence of an atomic attack . . . it began with a flash of light, followed by a ball of fire, a shock wave, and a huge mushroom cloud." This glimpse of the bomb, Sharp points out, was "firmly and unapologetically from the perspective of the bombers, the people who had deployed the weapon and watched its effect from a distance."[2]

The mushroom cloud remained distant from most American eyes through the end of the 1940s. During these years, American nuclear tests were conducted at the Pacific Proving Ground in the Marshall Islands. These tests proved expensive and increasingly difficult to hide from international scrutiny, leading the AEC to conclude a domestic test site was essential to remaining competitive in the global arms race. That Pacific nuclear tests were repeatedly displacing and heavily irradiating the residents of the Marshall Islands was not a significant factor in the decision to seek a domestic site. Southern Nevada seemed to fit the bill. Close to Los Alamos Laboratory, where much of the weapons design occurred, and on the edge of the Colorado Plateau, a major source for the uranium necessary for weapons development, the desert region just north of Las Vegas seemed like a wasteland to the rest of the country. In what has become an oft-quoted statement, then AEC chair Gordon Dean suggested the arid landscape was "a good place to dump used razor blades." In 1951, the Nevada Proving Ground, later known as the Nevada Test Site (NTS), was carved out of Nellis Air Force Base, which had been established during the early years of the Second World War.[3]

Technically, the military did not hold legal title to all the land it had appropriated. Seventeen different tribal groups belonging to three major ethnic groups—the Western Shoshones, the Southern Paiutes, and the Owens Valley Paiutes—had ancestral claims to the region. These peoples had actively hunted and gathered across the region prior to the government's appropriation of the land for the air force in 1940, and thousands of sacred sites, including petroglyphs and burials, were located and later would be tallied across the Nevada Test Site. Nearly seventy years after the military first claimed the land, much of the region still legally belongs to the Western Shoshone Nation, to whom it was deeded in the 1863 Treaty of Ruby Valley. Rebecca Solnit explains that the "federal government has admitted [Shoshone] legal ownership of the land to the extent of trying to force the Shoshone to accept payment for it, but the paltry sum allocated still sits in a Department of the Interior bank account." Like the Marshall Islanders, the Shoshone and Paiute peoples lacked the financial or political clout to influence the military's decision to rehearse nuclear war on their homelands, and bombs began to rain down regularly on American soil in January 1951.[4]

The inception of nuclear weapons testing in Nevada received little fanfare in local newspapers such as the *Iron County (Cedar City UT) Record*, and for the first two years of NTS operation, the test site's neighbors did not seem to find it threatening. Military leaders already believed the public harbored an "unhealthy, dangerous, and unjustifiable fear of atomic weapons," thus any indication of public danger from domestic testing was seen as a threat not only to the test site's operation also but to national security on the whole. Historian

Frederick Schoemehl documents how early AEC press releases offered few details about scheduled tests and did not give citizens cause to be concerned. Publicity emphasized "biographical sketches on senior Los Alamos and AEC officials" and suggested that only those people who wandered directly onto the bombing range would be in danger. Myron Glazer and Penina Glazer argue the military and government officials' suppression of accurate information in such cases is a fundamental condition of the national security state. They write, "Secrecy [becomes] a crucial component of many decisions and even well-informed residents [are] unable to learn much detail about activities in their own communities."[5]

Grazing their sheep north of the test site in the winter months, brothers McRae Bulloch and Kern Bulloch were aware of the tests but not cognizant that any danger existed. "We'd be there early in the morning and the sky'd light up, it'd just light up, before daylight," McRae recalled. "And then we'd see them shooting up big mushroom clouds in the day. You could see the fallout coming right up the valley. You could just see it. Just kind of a smoky cloud. Until this '53, we never had no trouble."[6]

The spring of 1953 marked the beginning of the Nevada Test Site's fourth test series, code-named Operation Upshot-Knothole. The first three Nevada test series—1951's Operation Ranger and Operation Buster-Jangle and 1952's Operation Tumbler-Snapper—drew little public attention. Emboldened, administrators planned more tests of larger devices for Upshot-Knothole. Beginning on 17 March and running through 4 June, Upshot-Knothole involved eleven bomb tests, or "shots." Six of them used devices that yielded between 20 and 61 kilotons of explosive power, with each kiloton being equal to a thousand tons of dynamite. For comparison, the bombs detonated over Hiroshima and Nagasaki in August 1945 were 12.5 kilotons and 22 kilotons; Upshot-Knothole tests Simon, Harry, and Climax would detonate with a force of 43 kilotons, 32 kilotons, and 61 kilotons, respectively.[7]

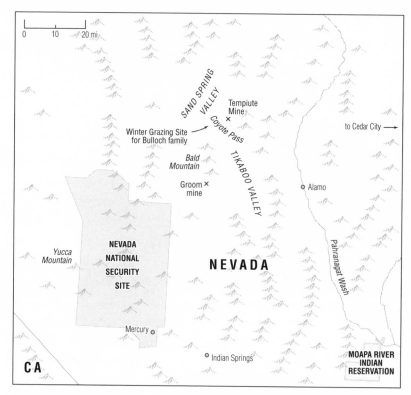

Map 3. Winter grazing site of the Bulloch family's sheep herd in 1953. Map by Erin Greb Cartography.

While the *Iron County Record* mentioned the nuclear tests occasionally, the photographs on the front pages of Cedar City's newspaper featured sheep more often than mushroom clouds. A sampling of the newspaper's reporting in the spring and summer months of 1953 is scattered with photographs of prizewinning rams and champion ewes, articles analyzing the shifting price of breeding sheep, advertisements for lectures on range nutrition, and announcements for the Wool Princess pageant. A 9 July photograph showed First Lady Mamie Eisenhower "admir[ing] a blanket made of Utah wool," with the caption "Mamie's little Lamb." The *Iron County Record* covered children's 4-H projects of raising lambs and women's activities in the active Wool Growers Auxiliary, an organization that represented women's expertise in the male-dominated industry. The auxiliary's "Make It Yourself with Wool Contest" announcements urged high school– and college-age women to participate in the wool industry by "mak[ing] their own suits, coats and dresses from virgin wool fabrics" in hopes of winning a Singer sewing machine or a $300 scholarship from Pendleton Woolen Mills. The *Iron County Record* also reported on social gatherings in the ranching community, such as the 30 July 1953 Wool Growers Auxiliary potluck supper. Attendees "enjoyed community singing" and "two skits . . . the first being a reading [of] 'The Death of a Sheep Herder.' . . . The second reading was a farce entitled 'No, No, a Thousand Times No,' by Mrs. Lehi Jones, Mrs. J. M. Palmer, Mrs. McRay [*sic*] Bulloch, Mrs. Wesley Bauer, and Mrs. Kern Bulloch."[8]

The *Iron County Record* article indicated a good time was had by all at the potluck, and it failed to allude to the somber topic likely under discussion that evening. A mysterious ailment had swept through several Iron County herds that spring, inexplicably killing off thousands of ewes and decimating the spring lamb crop. Sixteen local ranchers, including the Bulloch brothers, lost substantial portions of their herds. They represented "approximately half of the operators who operate[d] on the West Desert on the Utah-Nevada line or in Nevada, from Iron County," according to Iron County agricultural agent Stephen Brower's report.[9]

The losses devastated Cedar City's sheep industry. Ranchers took out loans, sold off land, and declared bankruptcy. Many eventually withdrew from the business. Several died young of heart attacks or cancer, and stories told about some ranchers' deaths connected the men's health problems with the trauma of losing their herds. Decades later, there is a potent irony to the potluck performances of "Death of a Sheepherder" and "No, No, a Thousand Times No."

Today sheep ranching has faded from visibility in Cedar City, and the catastrophic 1953 die-off is not a significant part of the community's identity. Local historian Janet Seegmiller, whose father-in-law lost "a third of his herd," suspects "the family just went through it, because they were always poor. Sheep didn't make a lot of money . . . so the boys I don't think ever thought about what they might have lost in terms of inheritance." While the impact of the deaths of roughly five thousand sheep may not resonate with later generations, at the time it proved a devastating blow—not only to the local economy but also to the spirit of a community whose identity was bound up in its ranching industry. It was also the first major incident downwind of the test site to shake the citizens' trust in the honesty and intentions of their government, which continued to maintain test site activities posed no danger, even when those downwind believed they could see the manifestations of that danger with their own eyes.[10]

As with many of the other ranching families who attended the potluck that July night, the Bulloch family's sheep operation was generations old and steeped in local experience. The Bullochs' ancestors were pioneers with the Church of Jesus Christ of Latter-Day Saints, some of the earliest whites to settle in Cedar City. In 2005 McRae Bulloch neatly summed up his family history in the region: "David Dunn Bulloch, my granddad, he [said], 'Let me go take a drink out of this crick so I can say I'm the first white boy that's drank.' So he run ahead, took a drink out of . . . Coal Crick. . . . And then they went in the livestock business. And it was just handed down to me, you know. Me and my brother."

By 1953 McRae and Kern, his younger brother, had assumed most of the responsibility for running what was still their father's sheep business. Decades of periodic drought and overgrazing left most of the rangeland in the vicinity of Cedar City too depleted to support local herds, so the Bullochs purchased a section of Nevada rangeland near Alamo, Nevada, approximately 150 miles west of Cedar City and just north of the newly commissioned Nevada Test Site. In October of every year, they herded their twenty-five hundred head slowly west, as the sheep browsed on the "white sage, bud sage, rice grass,

Brigham tea, black sage, [and] blue sage" and ate snow for water. Arriving on their Nevada range a month later, the Bullochs grazed the sheep until late March or early April, when they began trailing them back east to Cedar City to lamb.[11]

McRae Bulloch has recounted his story of the events of 1953 hundreds of times for other ranchers, for lawyers, for senators, for journalists. In decades of retelling, his story has varied little. In 2005, he related it for this author while sitting in his Cedar City living room beneath a scene of browsing sheep that his wife, Lillian, had painted. "In '53 we were there, me and my brother. Guys come up from the valley—they were army personnel—and they come up the valley there, and we were in Coyote Pass, that's between Tikaboo Valley and Sand Spring. We were there with our sheep. Drove up to camp in a jeep, several of them. And he said, 'Hey, you guys are in a helluva hot spot here with these sheep.' They'd set that bomb off; we'd seen it drop that morning. We watched it. They said, 'You're in a helluva hot spot with these sheep. You better get them out of here.'"

McRae and Kern could have ridden away from the area on their horses, but doing so would have necessitated abandoning their animals, which represented the entirety of their family's livelihood. Large herds of sheep move only a few miles each day and are even slower in the springtime when the ewes are heavy with lamb. The Bullochs lacked any way to transport all the animals, leaving them with no exit from what they now knew to be a "hot" area.

"We had no place to go, you know. We can't go any place," Bulloch explained. "And they radioed back to their company, I guess, their headquarters, and they said, 'There's a herd of sheep and two sheepherders up here in a helluva hot spot. What are we going to do about them?' And they said, 'Well, how many guys are there?' And they said, 'Well, there's just two, there's two men.'" McRae Bulloch heard the voice on the other end of the radio reply: "Well, you better go up on up to Tempiute Mine. There's 300 people up there." The brothers knew the mine was located several miles away. As they listened, the voice on the other end of the radio instructed: "Forget about the sheepherders and go up there and put those people at Tempiute under cover. Get them protected." The army personnel, who McRae describes as having "protective cover on their hands and on their feet and everything," repeated their warning to the ranchers: "'You guys better get out of here. You're in a hot spot. You're in a heckuva hot spot.' We couldn't go. I mean, we had no place to go."

The Bulloch brothers had no hard evidence to substantiate their story when they retold it to their families and fellow ranchers over the coming weeks, yet they were believed. While they did not have photographs of the mushroom cloud or Geiger counter readings of the radiation on the range that day, they had their reputations. They were local sons who carried the name and tradition of a well-respected ranching family with deep roots in the region. By noting their location, "Coyote Pass . . . between Tikaboo Valley and Sand Spring," they substantiated their story in an important way for local audiences. Folklorist Suzi Jones calls this technique regionalization. "Environments are loaded with meaning for their inhabitants," Jones observes, "from the very primal notion of 'home' which a simple place name can evoke to the elaborate cosmological schemata which have been imposed on landscapes." Evoking this shared meaning enables the narrator to engage others from the geographic area more fully in the story they are relating, lending a greater sense of authenticity and truth. When the Bullochs used place names to describe where their herd had been when they encountered the AEC personnel, they offered local listeners several layers of information. Those who had been to Coyote Pass or heard other stories about it found they had an emotional connection to the setting of the story they were being told. Any local with a rudimentary understanding of regional geography could also understand how close the ranchers had been to the test site and how close the test site was to places such as Bald Mountain and Tempiute Mine.[12]

McRae Bulloch's repetitive use of the phrases "we had no place to go" and "you're in a helluva [or 'heckuva'] hot spot" clearly expressed their desperate and untenable situation. As he recounted listening to the radio exchange, his audience realized, as he and his brother had, that "just two men" and their sheep would receive no protection from these military personnel. Describing the army men as having "protective cover on their hands and feet and everything" further underscored the Bullochs' vulnerability to radiation.

McRae and Kern Bullochs' story was later corroborated by other civilians—namely, the Sheahan family who operated the Groom Mine and lived roughly eight miles to the southwest on the south side of the Tempiute Mountain Range. Testifying in United States District Court several years after the incident, Daniel Robert Sheahan regionalized his recollections as McRae Bulloch had, pinning down his account with specifics about the locations in which events occurred. Shortly prior to the March tests, Sheahan recalled, a Public Health Service radiation monitor named William Holy came to the Groom Mine to set up a monitoring station "two hundred feet from [the family's] main cabin." Public Health Service monitors, who were stationed throughout the downwind region, worked closely with the AEC and rarely explained anything to the families in the area about the measurements they were taking.[13]

The Sheahans provided Holy with one of their cabins in which to set up his equipment. Sheahan testified that following a test on "March 24 . . . the cloud from the dust and so forth, headed north, just over the edge of our mining area and back towards the Bald Mountain." That morning, Sheahan walked down to the cabin where Holy was working to invite him to breakfast in the family home. "He stated that he couldn't leave his equipment; he had to be there, because he was expecting messages from the Tempiute side over the phone system." Later that day, Sheahan and his father spoke with Holy again. "As I remember it, Mr. Holy stated that the fallout had landed on a band of about 2,000 sheep in the Pennoyer Valley, and there were two sheep-herders with this band of sheep. That was all I heard about it at that time."[14]

Subsequently, according to Sheahan, Holy brought up the sheep again while eating supper with the family. "My father and Mr. Holy and I were discussing the fallout in the different areas, and something came up about the sheep. And Mr. Holy said, 'It doesn't look like they are going to do anything with this band of sheep.' And my father got kind of irritated. . . . He said, 'It's an awful thing that they will allow two sheepmen to remain in an area like that, when they wouldn't dare send the same material south, to Las Vegas.'" When asked how Holy responded to his father's statement, Sheahan testified that Holy replied: "I agree with you."[15]

Like other residents of the downwind region, the elder Mr. Sheahan had begun to apprehend a fundamental and unpleasant reality about the nuclear testing program taking place near his home: while local residents were assured they had nothing to fear from the tests, explosions occurred only when the wind was blowing due east, away from densely populated California and toward the more sparsely populated communities of eastern Nevada and Utah. This fact was not lost on downwind residents, many of whom made their living outside. They were cognizant of the wind patterns and within sight of the bright flashes and mushroom clouds. Initially, the Sheahans, like the Bullochs, took the government at its word when its representatives claimed nuclear testing posed no danger.

In the years after World War II, when the spirit of victorious nationalism ran high, many Americans freely and even fervently expressed their loyalty to the government. The Sheahan family, for example, provided working and living quarters for a government health monitor participating in the nuclear testing program and welcomed him to family meals. As the early years of the testing program wore on, the Sheahans joined others in the downwind region in second-guessing the AEC's assurances of safety. The government's apparent sacrifice of the two unknown sheepmen, their literal countrymen, gave credence to their apprehensions. As the Cold War heated up, unquestioning loyalty became harder to give, a contribution the government demanded rather than inspired. It was understood, and frequently spoken of in propaganda and the press, that the government was working very hard to keep Americans safe and that Americans ought to support their government in every way they could and not ask too many questions. Fears of communist infiltration led to witch hunts in workplaces and loyalty oaths in nearly all industries.

Meanwhile, with "nowhere to go," the Bulloch brothers continued to herd their sheep slowly back toward Cedar City for spring lambing. "Our sheep had to eat snow for water, that's what they did, and then they ate the grass, you know. But after that, why, we started to notice, skin scabs on their noses and scabs on their ears." Having been told they were in a hot spot may have preconditioned the brothers to look for oddities among their herd, but as McRae Bulloch recounted, so many strange symptoms began to occur it was impossible not to detect them. "We started noticing white spots coming on [black sheep's] wool, on the top of their backs. We had a black horse there at camp, and pretty soon he had white spots come on his back and on his rump and all over, from this fallout, I guess. We said, 'Well, goll, that must've been powerful stuff.'"

The scabs on the faces of the sheep and the white spots on the backs of the black animals were powerful indicators for the sheepmen. Years of experience running sheep in the region had familiarized them with the plants and conditions in the area, as well as the assorted ailments that sheep tended to develop while browsing there. Having never seen these sort of discolorations or facial scabs before, McRae and his brother connected the symptoms to the new condition in the region—the dust that had followed the atomic tests, drifting onto the animals backs and coming in contact with their faces as they grazed on local vegetation.

As they continued the slow return to Cedar City, the Bulloch brothers noticed "a lot of weak [sheep] that would kinda hang back. Funny, they would just die. A lot of our ewes aborted their lambs." Many of the sheep collapsed on the trail. The Bullochs resorted to loading them onto "a sixteen-foot grand cattle ride, two-and-a-half-ton truck." McRae explained, "The sheep would give out, we'd lift them on the truck, and bring them on and then haul them the line. By the time we got to the ranch we had it full. We couldn't bring any more sheep on it. And every sheep was dead that we put on that truck."

Asked if he recalled ewes aborting their lambs on the trail prior to that season, Bulloch responded, "It never happened before." He continued, "We got into our lambing shed here [in Cedar City], why, we started to lamb. The lambs were born dead, a lot of them. Not all of them but a lot of them were born dead and born with thin hide. You could see the part'd be kinda [thin] by the heart, not much wool. And they'd stand but then they were weak, they couldn't nurse. And we wondered what was the matter."

While they suspected the problems to be connected to radiation exposure on the Nevada range, the Bulloch family continued to exhaust every possible explanation, adjusting all the factors under their control. "We thought, 'Well, maybe we were giving them too much grain,'" McRae related. "We were graining them and had done out there [in Nevada]. We had a supplement of corn and barley and cottonseed and salt. We used to mix it all together. If we didn't want them eating too much we'd put quite a lot of salt, like maybe, a third. And then if we wanted them to eat more grain, why, we just put an eighth or less. We regulated how much they ate with the salt. If it's salty, they don't eat so much. We used to call it mush; we were feeding them mush and did all the time." Bulloch described using "a wagon with a dozen troughs, long troughs, in the wagon that we hauled our mush in, and then our water troughs. When there wasn't snow, we hauled water."

As he related the events of 1953, McRae Bulloch periodically deviated from the chronology to explain various aspects of the ranching industry, such as the inclusion of a few black sheep to assist in keeping track of many hundreds of white sheep or the way ranchers used salt to control the amount of grain the sheep consumed. His explanations were more than just a courtesy to audiences unfamiliar with the work of sheep ranching; they also confirmed his expertise as a rancher. Every additional detail Bulloch provided, from the dimensions of the cattle truck the brothers used to haul the ailing sheep to the specific deformities and weaknesses of the newborn lambs and the setup of the feeding trough wagon, shored up the audience's impression of his vast experience raising sheep.

After each explanation of the ranching industry, Bulloch returned to the events of 1953. "Getting back to here, in Cedar, when we were lambing, why, the lambs just started dying and some of the ewes." The ranchers called in local veterinarian Dr. A. C. Johnson. "But then he looked them over, and he said, 'Well, I don't know. I haven't got an answer for it.'" The Bullochs shared their frustrations with other ranchers in the area. "Finally, we got talking around, and there was other sheep men who was having the same problem: the Clarks and the Seegmillers and the Coreys. There was a whole bunch of them."

Consulted when the local veterinarian could not identify the ailment sweeping the herds, Agricultural Agent Stephen Brower found the symptoms confusing as well, but he suspected the nuclear testing might have been at fault. An employee of the Extension Service at Utah State Agricultural College (USAC), Brower served as a conduit between the county's farmers and ranchers and the agriculture experts at the college in Logan, communicating new research and techniques to the locals and relaying their problems, concerns, and conditions back to the college. At the inception of the nuclear testing program in Nevada, the AEC had contacted Brower in his capacity as county agricultural agent and "made a point of reassuring [him] that there was no problem." AEC officials had "encouraged [him] to reassure the local people that there would be no danger, that the levels of radiation would be negligible." The AEC also had provided Brower with a Geiger counter, which he used to "check vegetation primarily in the west desert and the Escalante Valley," areas in which Cedar City ranchers grazed their sheep herds. In doing so, Brower frequently discovered radiation at two to three times the normal background level on the plants the sheep foraged. Concerned that radiation might indeed explain the problems in the herds of the Bullochs and other local ranchers, Brower "made contact with the state vet, the agricultural experimentation station . . . suspecting it had something to do with the fallout but we had no experience."[16]

Several days later, on 24 May, F. H. Melvin, head of the U.S. Bureau of Animal Industry's Salt Lake City office, and John Curtis, Utah State veterinarian, inspected the Cedar City herds. Melvin reported that even at that early stage, ranchers were "very convinced that their sheep were suffering from the effects of atomic experiments, while they were in that area in Nevada." Utah commissioner of health George Spendlove became involved, and on 2 June he informed the AEC of the sheep deaths.[17]

Shortly thereafter scientists, radiation experts, military veterinarians, and AEC officials began descending on Cedar City's lambing sheds to investigate the losses. As McRae Bulloch pointed out, by the time these outsiders arrived, the most severely affected sheep were "just a pile of bones," a factor that handicapped the investigation significantly. Several months had passed since the herds had been on the Nevada range, complicating efforts to reconstruct the radiation doses the herds might have received. Despite the decay of the carcasses, McRae Bulloch recalled, one investigator "put the Geiger counter over [the bones]. One of us said, 'Is it hot?' And the guy that had the Geiger said, 'Hot? It's so damn hot that this needle goes over so fast it just about jumps the pole!" The AEC later denied that statement. It stuck in the minds of the ranchers, though, and Bulloch and others repeated the anecdote many times.

The AEC's internal agenda compromised their investigation and response. The Upshot-Knothole test series had already attracted unfavorable publicity, and AEC commissioners perceived a "serious psychological problem" of public concern developing over testing safety. Several springtime shots (including Simon and Harry) dispersed considerably more fallout outside of the test site than scientists deemed acceptable. Local residents were asked to hose down their cars on multiple occasions, and the detonation of Shot Harry (later known as Dirty Harry) on 19 May caused so much off-site fallout that the AEC advised residents of St. George, south of Cedar City, to stay indoors for several hours. In the wake of these tests, and the livestock deaths and complaints of human illness that ensued, AEC commissioner Eugene Zuckert privately concluded, "In the present frame of mind of the public, it would take only a single illogical and unforeseeable incident to preclude holding any future tests in the United States."[18]

The possible preclusion of nuclear testing in the United States was of serious concern to military leaders and politicians, who believed staying ahead in the global nuclear arms race was more important than ever. Fears of international and domestic communism ran high. As southern Utahans hosed off their cars after Shot Harry, that May the *Deseret (Salt Lake City) News* ran headlines such as "World-Menacing Russ[ian] Atom Pile Seen within Year," "UN Troops Reel Back under Fierce Assault" in Korea, and accused "A-Spies [Julius Rosenberg and Ethel Rosenberg] . . . Hear Death Date for Fourth Time."[19]

The massive sheep die-off presented precisely the sort of incident AEC commissioners feared. "At the time the first team of investigators arrived from Los Alamos," Stephen Brower recalled, "the head of the biological medicine division of the AEC . . . Dr. [Paul] Pearson, told me . . . that the AEC could under no circumstance allow the precedent to be established in court that the AEC was liable for radiation damages, either to animals or humans." Just as it had hired the *New York Times*' William Laurence so that it might influence American perceptions of nuclear weapons, the AEC now sought to influence the scientists and veterinarians tasked with explaining the sheep deaths. This manipulation constituted a calculated effort to shape the public's perception of nuclear weapons development and preserve the country's edge in the arms race. Like the Bulloch brothers, the AEC was bent on telling a story that supported its cause. Since all the evidence was on the side of the sheep ranchers, the AEC used its considerable influence to craft an official fiction in order to maintain the illusion that nuclear testing did not pose a risk to the downwind region.[20]

Internally, government scientists confirmed that radiation "was at least a contributing factor to the loss of these animals." In his initial report to the Atomic Energy Commission, dated 17 June 1953, army veterinarian Maj. Robert Veenstra confirmed that the thyroid regions of numerous sheep produced high readings on a Geiger counter and bore lesions "typical of the 'fall out' lesions in the Trinity cattle." Further, he suggested, "the location of the lesions and the nature of the sheep to nibble grass short, leads one to suspect that the lips and fore face could easily come in contact with material on the bushes . . . that would cause these lesions."[21]

As Veenstra's reference to Trinity cattle (exposed to the 1945 nuclear test Trinity in New Mexico) suggests, the AEC had conducted multiple studies of radiation's effects on livestock prior to 1953. Historian Howard Ball notes that as early as 1951, "AEC reports warn[ed] that years after the 1945 Trinity atomic test, cattle and sheep were still showing the effects of radiation burns and of having ingested radioactive vegetation, which led to deformities in the animals' offspring."[22]

In addition to its own radiation studies, the AEC also funded studies of radiation at colleges and universities, including Utah State Agricultural College, the center of agricultural knowledge in the downwind region. In June 1950, the AEC offered USAC a cooperative contract. Acknowledging "the College has conducted basic research relating to atomic energy as part of its regular program on its own initiative," the AEC expressed its "desires to encourage, further, and intensify this basic research work in order to increase fundamental knowledge which is essential to progress in all aspects of atomic energy development." The contract established a "simplified lump-sum" payment arrangement "under which the Commission [could] cooperate with the College in intensifying this basic research related to atomic energy." Further correspondence from the AEC requested all research results be submitted to the commission for screening before USAC released them to the public. College of Agriculture dean R. H. Walker acquiesced to the secrecy agreement, replying, "We shall be happy indeed to cooperate as you have suggested."[23]

While none of USAC's research projects prior to 1953 appear to have focused on the effect of radiation on sheep, the financial relationship between the college and the AEC—and the secrecy agreement it entailed—raise doubts about the ability of USAC's extension service to fulfill its stated commitment to "respond to critical and emerging [agricultural] issues with research-based, unbiased information." Even though most ranchers were likely unaware of the college's connection to the AEC, some drew their own conclusions when USAC's agricultural extension service failed to explain suspicious livestock deaths.[24]

Logan rancher Willard "Bill" Peterson brought dead animals to the agricultural college in the mid-1950s for autopsies after three hundred of his ewes died one night on the range and again after a wave of mortality in his spring lamb crop. "They said they didn't know what it was. Well, they're funded by the government, and so I don't imagine they're going to be too interested in telling me it was radiation, when their research and all their finances depend on it." Marjorie Black ranched north of the west-central Utah community of Delta, on land her husband's family first settled in the 1880s. One day in the 1950s, she and her husband discovered twelve of their cattle dead. "The strange thing was, and that's why I contend it was concentrated fallout from the bomb testing in that one area, because outside the fence the cows were fine. Inside, not only were the cows dead, but there were rabbits and magpies dead," Black told writer and photographer Carole Gallagher in 1988. "I called the county agricultural agent in Nephi, Ray Burtonshaw [sic]. . . . He called Dr. Don Thomas from the Agricultural College in Logan and they both came out. I went down into the field where the cows had died with them and said, 'Can you check for fallout?'" According to Black, the agricultural agent and the professor replied, "All we're going to do is check the water. Maybe the water is poisoned." With certainty, Black retorted, "It isn't the water. This has been a water hole for the cattle for years and it isn't the water." The men replied, "Well, we can't check for anything else." There is no evidence that field agents Ray Burtenshaw, Stephen Brower, or Dr. Don Thomas were influenced by their employer's connection to the AEC. Burtenshaw, Thomas, and those USAC experts who also examined the carcasses of Bill Peterson's sheep may have simply believed they had no way to assess radiation-related death or the presence of fallout. Many did have Geiger counters issued to them by the AEC, though, just as Stephen Brower's had been.[25]

It is easy to see how locals began to suspect a cover-up, particularly as rumors about fallout and ailments seemingly connected to the test site continued to swirl across the countryside while government officials continued to deny there was any risk. Not long after the death of her cattle, Marjorie Black experienced a severe, burning headache that lasted several days. Her doctor suggested she contact officials at the Dugway Proving Ground, where chemical weapons testing was occurring to the west. "A fellow came out, they landed the helicopter down below the corrals. I told them about the cows and asked if they'd go down, and they wouldn't go down into the area." Later when Black pressed Dugway officials for more information, she says they denied ever visiting her ranch. She lived close enough to the test site to see the mushroom clouds rising over southern Nevada. "I saw when they set some of those bombs off one morning. I was by the window, looking to the southwest, when it went up like a fan of colors. It spread way out, and then later it shook the windows in the house until I thought it was going to break them." Black remembered how her daughter's friends who lived "west of Gold Hill, west of Ibapah, close to the Nevada line," experienced acute health problems after the tests. Her daughter "stopped out to see this lady, and she had these burns on her face. [My daughter] asked what was the matter. . . . She said, 'Oh, these are burns from the bomb testing. You ought to see our cattle.' The hair was burned off in strips. Her husband was out riding to check the cattle, and within a few hours he just dropped off his horse dead." Black reflected, "It was amazing how you come up against a blank wall when you go to them for information on what was going on. We had no answers. They said a lot of things were harmless when they weren't."[26]

As the AEC investigation into the Cedar City sheep die-off got underway, Agricultural Agent Brower became convinced the AEC was being less than forthright about its findings. After accompanying veterinarians R. E. Thompsett and Veenstra to inspect the Cedar City sheep, Brower asked them to provide a copy of their findings, and "they agreed to do [so]. . . . [Veenstra] later reported that he was unable to fulfill the promise because all copies of his original report, including his file copy, had been taken by his superior at AEC and he had been instructed to rewrite the report and delete any reference to his diagnosis of the symptoms as being radiation related."[27]

As the Cedar City investigation progressed, AEC representatives began publicly discounting radiation as a cause and suggesting malnutrition was a more likely culprit. Other scientists, especially those connected with public health organizations, remained unconvinced. Locals did not find the explanation satisfying at all, as the 2 July *Iron County (Cedar City UT) Record* indicated. Under the headline "'Malnutrition' Diagnosis Resented; Authoritative Report Expected Soon," the paper reiterated that the disease had never been seen before and assured readers that "local livestock men and interested residents will not dismiss the matter as a simple case of malnutrition."[28]

On 15 October, sheep rancher and well-respected city councilman Doug Clark became involved in a confrontation with the AEC's Lt. Col. John H. Rust. Stephen Brower, who witnessed the exchange, recounted it years later in a letter to Utah governor Scott Matheson:

"Doug asked for some information on the similarity of the conditions of their sheep and behavior of research animals. Somehow he had gotten some confidential data on this matter from other veterinarians. The colonel attacked him verbally about him being a dumb sheepman and not being smart enough to understand it if he did give him the data." Brower watched as "for ten or fifteen minutes the colonel harangued, belittled, and discredited Doug's intellectual ability to understand anything. In less than two hours after this incident Mr. Clark was dead of a heart attack."[29]

Word of Clark's death, and the confrontation that preceded it, spread rapidly among the sheep men. In the retellings, minor details of the story shifted, and individuals added their own rhetorical flourishes and commentary though the language, the story line, and sentiment remained the same. Rancher Delayne Evans, who was not present at the confrontation, recounted the incident to Carole Gallagher to illustrate the injustice of the investigation. "Now Doug Clark, the Atomic Energy guys visited him one day and called him a liar until he had a heart attack and died right on the spot. They told him, 'You're just dumb, you don't know what you're talking about.'" On the contrary, Evans asserted, "he knew just exactly what he was talking about, he was the County Commissioner at the time and an outstanding citizen. We knew it caused it, hell, we're not stupid. They thought we were a bunch of stupid sheepherders but there's nothing stupid about us."[30]

McRae Bulloch, who remembered Clark as "a city councilman and a very prominent and bright man . . . no stupid man," surmised that stress over the debt from his losses had played into Clark's death, just as similar losses had affected his own father's health. Bulloch said, "Doug Clark, now he got after them; they took some scientists out there. And he was so worried, he [was] in debt, and that's the way my father was too. He was in debt, they were both in debt, and we had to pay our bills, you know." Bulloch was not present at the confrontation between Clark and Lieutenant Colonel Rust, but he described the same version of events: "Doug Clark, he had a row . . . an argument, and they said it wasn't their fault. They called him a dumb sheepherder, he didn't know what he was talking about. He had a heart attack that day and died. . . . You heard about him. Well, he died, and that was the reason. He was just in debt; he didn't know what he was going to do. He was losing his sheep."[31]

The story surrounding Clark's death also spread quickly to the broader community. In later years, the incident became an oft-cited illustration of what many downwind residents began to see as the dismissal of their region by federal officials. Loa Johnson was a mother and music teacher in Cedar City at the time of the sheep die-off. In 2005 she recounted the story of Doug Clark's death. Like Bulloch and other downwinder narrators, Johnson prefaced her description of Clark's death with an acknowledgment of his respected standing in the community: "One of the very big sheep men [Doug Clark], . . . he had gone on to get graduate degrees. Very intelligent man. But he loved his ranching, and that's all he wanted to do." She explained that Clark "was talking to one of the military personnel. I don't know whether he was a general or just a major or colonel or something. He was fairly high up. He called [Doug] an ignorant farmer and [said,] 'You don't know what's going on and how can you know what's going to affect your sheep and your town?' . . . That was their attitude," Johnson explained. "The people that lived here were nothing but ignorant people who didn't know anything and shouldn't be worried about it. Well, [Doug Clark] was a much more refined kind of person than the military man he was talking to, so he walked off and he was so angry that he walked right over to his house trailer and had a heart attack and died."[32]

Clark's prominence in the community and the circumstances preceding his death made him a powerfully sympathetic character in downwinder stories. He came to symbolize the betrayed integrity of an honorable and patriotic community steeped in sheep-raising expertise. That Clark possessed a college education, as Loa Johnson indicates, but still chose to run sheep only added dignity to the ranching industry. Narrators recounting his character use such descriptions as "outstanding citizen," "bright," "very intelligent," "prominent," and "refined." These same narrators emphasize heavily that outside scientists and government officials saw the local community as "stupid," "dumb," and "ignorant," or the extreme opposite of the locals' perception of their own community as personified in the character of Doug Clark.

The AEC's final report, issued on 6 January 1954, concluded: "Abnormal losses suffered by the sheepmen can not at this time be accounted for by radiation or attributed to the atomic tests conducted at the Nevada Proving Grounds." The following week, AEC officials held a meeting with ranchers at the Cedar City Fire Hall. Stephen Brower introduced the ranchers to Dr. Paul Pearson and Col. Bernard Trum of the AEC and Dr. James Terrill of the Public Health Service. Together they made the AEC's case against radiation as a factor in the sheep deaths. The official transcript of the meeting is revealing. The ranchers in attendance asked pointed, practical questions about radiation exposure based on their daily observations of the sheep's foraging habits and symptoms. AEC officials responded to most questions evasively, making vague or complicated references to laboratory studies and voicing emphatic, often patronizing denials of excessive radiation. They avoided other questions entirely.[33]

Having been informed that radiation had not contributed to the deaths of the sheep, McRae Bulloch mentioned the Stewarts, cattle ranchers who shared water rights with the Bullochs on the Nevada range and who had recently been compensated by the AEC for radiation damage to their horses. He pointed out, "Those Steward [sic] horses are right next to us . . . the [army personnel] . . . told us we were in a hot spot." Dr. Terrill sidestepped Bulloch's statements, saying, "This has been a wonderful opportunity for us to evaluate the actual damage done with the known fall-out. You can't cause a big amount of damage with a small amount of radiation. . . . People get excited . . . we must realize they were measuring very small amounts of radiation." The AEC's Dr. Pearson chimed in: "Radiation has a lot of potential beneficial uses."[34]

Rancher Douglas Corry queried the scientists on details of AEC radiation experiments conducted on cattle. Upon being informed by the AEC's Colonel Trum that the cattle received radiation via either injection or capsules, Corry asked, "Where a sheep is getting dust on feed over [a] period of time how important is the accumulative effect?" Trum responded, "It is possible we can't get the dust but we know how much is in the dust and if we find any one thing we know how much of the others was in that dust. . . . In tests we have used far greater amounts than anything we picked up." Unsatisfied, the ranchers continued to pursue the question of gradual ingestion of radiation, rephrasing Corry's question at least three times, but each attempt received evasive replies.

Unimpressed by charts, calculations, and statistics about external gamma radiation, the ranchers drew on common sense and local knowledge in formulating their conclusion about the sheep deaths. While on the range, they witnessed the initial blasts of weapons tests and the clouds of dust that quickly followed. Dust clouds, in and of themselves, would have been unremarkable. As rancher Bill Peterson recalled, laughing, "Out in that desert, there's *dust!*" Ordinary dust, however, did not leave white spots on the hides and wool of black animals. Ranchers noticed the dust settling on the sage the sheep foraged on and the sores that appeared on the animals' faces soon after eating. In a business where success and economic viability depended on the thriving of healthy animals, the ranchers could ill afford to ignore what the sheep consumed.

Common sense tends to be undervalued as a source of wisdom, yet the body of knowledge it encompasses is by no means simplistic or generic. Anthropologist Clifford Geertz points out common sense is "as much an interpretation of the immediacies of experience, a gloss on them, as are myth, painting, epistemology . . . it is, like them, historically constructed and, like them, subject to historically defined standards of judgment." Common sense can be "questioned, disputed, affirmed, developed, formalized, contemplated, even taught, and it can vary dramatically from one people to the next. It is, in short," Geertz concludes, "a cultural system . . . and it rests on the same basis that any other such system rests; the conviction by those whose possession it is of its value and validity." Cedar City residents respected the titles and credentials of military leaders and government scientists, yet their trust of these outsiders was based on ideas—such as patriotism and admiration for formal education—rather than on experience. When the statements of these outsiders clashed with the commonsense opinion of respected local experts, experience won out. Patriotism did not put food on the table, and formal education could not fully acquaint one with the challenges of raising livestock.[35]

Decades later, it became apparent the Cedar City ranchers, in fact, had pinpointed the cause of the malady. In 1979 Dr. Harold Knapp, formerly of the AEC's Fallout Studies division, looked into the documentation from the 1953 investigation. His inquiry grew into a report, based on dose reconstruction estimates, that concluded "the simplest explanation of the primary cause of death, not considered at the time, is irradiation of the sheep's gastrointestinal tract by all the fission products present in the fallout particles which were ingested along with open range forage." Lacking Knapp's scientific credentials and access to government documentation, the ranchers who attended the 1954 fire house meeting could only continue to reiterate their observations.[36]

At the meeting a rancher identified only as Clark, likely one of Doug Clark's sons, pointed out, "You stated that you fed your control sheep many times what our sheep could have had. All our evidence [was] dead some months before any tests were made here. Your tests show radioactivity decreases. How can you know [how much our sheep received]?" Terrill responded with a convoluted explanation of calculations and concluded, "I don't think even if you were

riding the bomb two feet from the explosion it would kill them. . . .
I have given [a] sheep such strong doses that he was so hot I had a
30-minute time limit to take samples. He gained weight." This boast-
ful and condescending claim from a government scientist surely
rankled the ranchers in attendance who firmly believed radiological
exposure was at the root of their losses.

Clark pressed the scientists further: "We know that practically
all the sheep that range in that area had these effects. We fed these
sheep corn and tried to keep them up. I couldn't keep my sheep up
where they were able to raise a lamb. I had never seen it before. . . .
Starting in March." When Leo K. Bustad, a General Electric Company researcher brought in to consult on the sheep deaths from the
AEC's labs at Hanford, observed, "There is very little protein in corn
and they could be low in protein," Clark responded, "What about . . .
native feed[?] Sage is very high in protein." The transcript indicates
none of the scientists responded directly to Clark's statement.

AEC representatives and government scientists continued to insinuate malnutrition or some unknown disease had been at fault for
the animals' deaths. Pearson closed the meeting by reassuring the
ranchers of the AEC's "continuing interest" in discovering the die-
off's cause. "If this shows up another year we will be in a much better
condition to study this. . . . We would like to sponsor and conduct
some studies aside from radiation and to this end we are setting up
projects and providing some funds for the college whose business
it is to study this." He concluded by admitting that "we can keep no
one from putting in a claim. You could bring suit, but the AEC has
no authority for paying more than $1,000, and claims would have
to be judged on the basis of these studies." Despite Pearson's stated
inability to block a lawsuit by the ranchers, if the AEC could not
settle claims beyond a certain amount and if those claims would be
judged on the basis of the AEC's own classified data, there seemed
little chance that a lawsuit would succeed.

Three months later, "in cooperation" with the AEC, USAC's Agricultural

Extension Station began its study on the "effect of nutrition on pathology and productivity of range sheep." The study's first progress report, which spanned 15 April 1954 to 15 January 1955, failed to explain the losses of 1953. It did note that "lack of interest on the part of some sheep owners also made the work difficult." This brief mention of the ranchers' "lack of interest" in the study, almost an aside in the report, speaks volumes. None of the ranchers expected a study of range nutrition to explain their losses; for them, the study was an insult atop of injury.[37]

The years following the 1953 die-off plunged most of the Cedar City ranchers into even deeper financial straits. Delayne Evans recalled, "I bet there wasn't one sheep that was out there in '53 that didn't die within two years because I was buying sheep all the time to replace them. . . . It breaks you. It broke me. I had to mortgage my home to buy more sheep to start out again. It took me 15 years to pay it off." The negative economic effect of such major losses accumulated. As Annie Corry observed, "It takes quite a long time to build up a herd again. We just borrowed the money and when my husband died we were still paying interest on it. You're always in debt. It drove quite a few people out of the livestock business."

Faced with the loss of their livelihood and convinced the AEC's investigation had not treated them fairly, eight families chose to file suit against the federal government in February 1955. They were not seeking large damages, only compensation for the market value of their lost stock and the extra help and feed used during the crisis. The Bulloch complaint, the outcome of which would determine the other ranchers' cases, listed 1953 losses of "250 ewes after shearing at $25.00 ($6,250), 95 ewes before shearing at $30.00 ($2,850), 1262 lambs at $15.00 ($18,930), Extra help and feed $1,650"; and 1954 losses as "300 lambs at $15.00 ($4,500)" for total damages of $34,180.[38]

Suing the federal government in 1955 took conviction. Riding a wave of national panic over the perceived threat of communism, Senator Joe McCarthy (R-WI) and other national leaders had spent the late 1940s and early 1950s energetically impugning the patriotism of citizens who questioned the government's actions. Loyalty oaths became common conditions of employment in government offices, corporations, even schools; by the year the Bulloch case went to trial, more than nine million American workers had been vetted by loyalty security programs. Those who refused to take loyalty oaths or were suspected of membership in the Communist Party or other groups considered subversive found themselves out of a job. By 1957, Gerald Rosenberg notes, "approximately 11,500 Americans had lost their jobs under loyalty-security programs." For those who did wish to debate the country's politics, Rosenberg argues, "the right to participate in political debate . . . was protected only so long as they communicated 'acceptable' ideas. When they did not, the First Amendment and the courts did not protect their rights."[39]

Dissent had been cast as the equivalent of disloyalty. Some propaganda even went so far as to imply that questioning the nation's nuclear weapons development program was tantamount to inviting nuclear attack from the Soviet Union. This implication was made clear in the 1955 AEC film *Atomic Tests in Nevada: The Story of AEC's Continental Proving Ground*, which was produced to calm downwind fears after the unfavorable publicity of the Upshot-Knothole series. In one scene in the film, two men argue, presumably over testing, while behind them a civil defense poster depicts a nuclear warhead descending and the tagline "Enemy Target #1 . . . Civilians." A stern but friendly voiced narrator intoned that testing in Nevada was a matter of "national survival" and "essential to our existence as a nation" and reminded viewers that "the towering cloud of the atomic age is a symbol of strength, of defense, of security for freedom-loving people everywhere, people who want peace."[40]

The Bullochs and the other ranching families read the same newspaper reports and magazines the rest of the country did. They had surely come across William Laurence's Pulitzer Prize–winning description of a nuclear explosion and read about the godlike qualities with which he had imbued the bomb. They were concerned about nuclear secrets being stolen by the Russians, and they worried about enemy warheads threatening their families, as did Americans in the rest of the country. At the end of the day, however, they could not escape the fact that their livelihoods had been shattered and that their experience did not align with the story the government representatives were telling.

Given that most of the ranchers in the lawsuit were also members of the Latter-Day Saints church, their decision to proceed with a lawsuit alleging government dishonesty is all the more remarkable. "It takes courage for a Mormon to dissent," writer Wallace Stegner reflected in his history of Utah's LDS community. "Their whole habit and training of life predisposes them to obedience." Stegner's assessment seems harsh, but there is no question the LDS church aligned itself strongly with the federal government during much of the Cold War years and systemically discouraged church members from questioning the actions of church or government leadership.[41]

Darlene Phillips was a young LDS mother during the atmospheric-testing era. She told Carole Gallagher, "I was brought up with a notion of blind obedience." After consulting her church leaders for "some guidance about the issue of fallout, . . . [I was] told 'We're just good citizens and we're lucky to have these tests near us and we should be honored. This is our chance to prove that we are loyal citizens of the United States.'" Decades later, she reflected, "I think Utah is the one state where they could have done this," implying that Utah's large LDS population was unlikely to ask too many questions. Phillips remembered "some demonstrations when milk was becoming perceived as [contaminated by nuclear testing] in the late fifties, early sixties. . . . My girlfriend, who has never been Mormon, asked me if I wanted to go to the demonstration and take our kids and make an issue of it." Again, Phillips sought guidance from her church. "I asked my bishop what was the proper role for a good Mormon woman to take. He said 'No, you stay away from it, those people are Communists,' so I didn't go. To Mormons, the first law of God is obedience."[42]

Years later, most of the ranchers involved in the lawsuit continued to characterize themselves as patriotic Americans, and many remained dedicated members of the LDS church. Historian Schoemehl has suggested the "underlying momentum" in their pursuit of compensation for the sheep deaths issued "from the historical memory of a group who saw themselves as a persecuted people." It is true that in the LDS church a strong collective memory exists of persecution, perpetuated by both the U.S. government and non-LDS Americans, and that many western American ranchers resent the actions and regulations of the federal government, particularly when it comes to the management of public lands. In spite of these attitudes, there is little evidence that the LDS sheep ranchers believed the government dismissed them because of their faith. The ranchers, like most downwinders, came to fault the government for its dishonesty, for not respecting their knowledge, and for treating them as disposable, but they almost always referenced this treatment as the betrayal of a regional population, not a religious one. "I don't think there was a conspiracy against Mormons," St. George, Utah, resident Michelle Thomas mused in 2006. "I just think we happened to be living in Utah and we were a small population, and we were good sheep, and we owned good sheep, as it turned out! I just don't think that they targeted Mormons."[43]

By their own accounts, the ranchers' motive in filing the lawsuit was not out of simmering historical resentment over the treatment of Mormons, nor was it some sort of subversive antiwar action. The ranchers were self-described patriots with no interest in interrupting testing activity at the NTS. Their lawyer Dan Bushnell made this point clear in his closing argument: "Progress in atomic research is of great importance. With this we agree. But in our race to excel, to exceed and progress, and to be strong, it is my belief that we must not violate rights and concepts which have made us as a nation strong."[44]

The government's claim that malnutrition and mismanagement had caused the sheep deaths was an affront to the sheep mens' generations-old ranching knowledge, and their lawsuit was as much an assertion of their expertise as it was a demand for compensation. Ranching was more than their livelihood; it was a culture at the heart of their community life, one that had been handed down to them and one that they had planned to hand down to their children. While the patriarch of the Bulloch family, David Bulloch, is listed as the plaintiff in court papers, his sons, McRae and Kern, were the ones who testified in court. Wracked with depression and worry over the demise of his herds, David was in failing health and would not live to see the outcome of the lawsuit.

The trial proceeded in much the way as the fire house meeting had the previous year. The AEC employed or influenced experts on radiation, and the documentation from the 1953 investigation remained classified. Lawyers representing the government frequently invoked the priorities and imperatives of national security, and they tossed around scientific jargon that the ranchers had no way of refuting. Lacking access to their own expert witnesses, the ranchers could only reiterate their own observations. Annie Corry described the trial to Carole Gallagher: "It was pretty bad. The government people poked fun at [our] veterinarian, said he didn't know anything. They made it look like he was very stupid. They just figured we were a bunch of farmers, which we were. It was something *we* did to the sheep, something *we* fed them, *we* hadn't taken care of them, the range was dry that year. Of course, we knew what happened because running sheep your whole life, it was something that [had] never happened before and it's never happened since."

Decades later, after declassification, the documents suggested the AEC conducted itself in an "intentionally false and deceptive" manner, "pressure[d] witnesses not to testify as to their real opinions," and withheld "a vital report" while presenting other information "in such a manner as to be deceitful, misleading, or only half-true." None of this information was available to the plaintiffs at the 1955–56 trial, and on 26 October 1956, Judge Sherman Christensen dismissed the case on the merits. He conceded to the government's claims "that the losses suffered by the Bullochs were the natural result of unprecedented cold weather during the lambing and shearing of sheep, inadequate feeding, unfavorable winter range conditions, and infectious diseases of various types, the combined effects of which were thought by some during the preliminary investigation to comprise irradiation syndromes."[45]

The verdict severely rankled the ranching community and convinced few residents from southern Utah that radiation was not to blame. Sheldon Johnson, a World War II veteran who worked as an optometrist in St. George, an hour south of Cedar City, recalled that the "mismanagement" verdict "was baloney. I don't think you could have found a soul here that swallowed that, [not] one ounce. That's baloney. We know sheep, and sheep get along, and if they don't, they move out. No, no, . . . on that sheep case, that was a bunch of crooks." Johnson's statement is even more meaningful when contextualized within his perspective on testing: he felt many American lives were saved because the United States used nuclear weapons against Japan, and he was generally in support of additional weapons development and testing. Said Johnson, "My perspective of the atomic bomb is that it was a great blessing, and we need to watch it and take care of it, and I think we have done a very good job, [we have made] some mistakes, and some of those mistakes was the bomb tests that were out in the open that they didn't know it would radiate that much." While he attributes his first wife's fatal cancer, his own cancer, his son's Down syndrome, and the illnesses of many close friends, relatives, and neighbors to testing, Johnson does not condemn the government for downwind human exposure: "I'm . . . not accusing them of purposely trying to do something bad, because I don't think they did. I think it was pure lack of knowledge and lack of understanding of what was happening." Regarding the sheep,

Johnson stated, "I would be tough on them for covering up that, personally, even tougher than for us, even though we were humans and they were sheep. It was so obvious that that was a big mistake and they wouldn't admit it."[46]

Shortly before the trial began, the AEC briefly acknowledged livestock radiation injury in its film *Atomic Tests in Nevada*. The narrator said, "Yes, livestock, grazing within a few miles of the site of detonation have in a few instances suffered skin and eye injuries from radiation, but otherwise [they] were in good health. Justified claims by owners have been compensated. Extensive studies have concluded that animals grazing at greater distances have not been injured by radiation." The narrator reassured audiences that "the AEC will continue to give advanced notice [of tests] to livestock owners." When asked in 2005 if he or his fellow ranchers ever received advanced notice of tests, McRae Bulloch replied emphatically, "Never. Not *one* time. The only time they ever come was to warn us we were in a hot spot [in 1953, prior to the sheep die-off]."[47]

Nearly five thousand sheep died following the spring of 1953. The emotional impact of this die-off on the small community of Cedar City and the surrounding agricultural region should not be underestimated. That the deaths occurred at lambing time only amplified the pathos of the experience for locals. Many ranchers' lambing sheds were close to town, and neighbors recalled vividly the "two-headed lambs, and . . . piles of dead lambs" at a time of year when healthy newborn lambs ordinarily would have delighted children and passersby. Prior to the die-off, rumors had circulated about people's headaches and skin rashes and affected livestock following bomb tests, but these rumors lacked the substance, magnitude, and clear connection to testing that locals saw in the sheep's deaths.[48]

Decades later when asked to recall the moment at which they first began to question the safety of the weapons tests, many downwind residents pinpoint the sheep die-off. As McRae Bulloch recalled, "The sheep was first and then people got to thinking, 'Well, goll!

Maybe that's what killed my brother, and my dad!'" The ranchers themselves proved something of a literal connection between the sheep deaths and human ailments. "A lot of those sheep men died of cancer," Bulloch related. "Grant Lavins was out there where we were, and he died. He was one that was in the lawsuit. And Elson Webster, he died of cancer. And I think Seegmiller died with cancer. A number of them just died of cancer."

"Oh heck," Bulloch exclaimed, "I've had several. I had kidney cancer. . . . And I had prostate cancer, and I had cancer cut off my nose. If you get up close enough you can see it. One cut off my cheek. One cut off my lip. And a lot of them on my arms."

Annie Corry told Carole Gallagher, "I wondered if all the livestock men died of a heart attack like my husband. There were two of them just up the street that passed away sudden just like he did." The widow had herself been diagnosed with a blood disease. "I got down to about 85 pounds. It's an infection of the lining of your arteries, quite a rare thing. The first thing you do is lose weight and get anemic. I was like those old ewes, I just wasn't hungry, couldn't seem to eat." Like others in the ranching industry, Corry also voiced concern about the possibility that radiation had passed from the bodies of the sheep to other laborers and then to people working and shopping where the meat and wool were sold. "The thing that bothered me was when the wool was taken off and put into those bags. Have you ever seen men tromp wool?" After the sheep had been shorn, laborers climbed inside large bags of wool and, using their body weight, tromped the fleeces down to pack in the maximum amount for shipment. "What do you suppose happened to those fellows who was tromping that wool that was full of radiation?" Corry wondered aloud. "I bet they all died. I've often thought about them. That wool was packed and sold as usual, and look at the people that handled it, people who scoured it. It had to go through quite a few hands."

The livestock deaths witnessed by the Cedar City ranching community had parallels elsewhere in the region. In numerous other instances the bodies of range animals told their owners the story of invisible contaminants present in their environment. While not every owner was able to deduce the cause of their animals' maladies at the time, revelations that emerged years later about the extent of Cold War pollution have brought those accounts back to the surface and have enabled those who tell them to piece together fuller narratives.

Bill Peterson, the sheep rancher from Logan, wondered for years if the losses his family incurred in the early 1950s were due to nuclear testing. He too lost hundreds of sheep to an inexplicable malady and watched helplessly as subsequent spring lambing resulted in massive mortality, yet his herd wintered hundreds of miles north of the NTS and the Cedar City herds. It is certainly possible, though, that Peterson and his outfit were exposed to radiation on their northern Nevada range. In a 1952 weather report meteorologists Robert List and Lester Machta noted the "channeling effect of the mountains north of the test site" that funneled fallout clouds rapidly north toward Elko, Nevada, in the proximity of the rangeland the Petersons utilized.[49]

The symptoms Peterson noted in his herd were inconsistent with those of the southern Utah herds, although Peterson likely would not have known that at the time. Checking on his herd in the Hansen Valley in western Utah, near the Nevada border, Peterson discovered "dead sheep all over the place when the sun came up. . . . There was over 300 that were dead. And their heads were puffed up, and irritated, looked like it swelled up, and red and stuff." In contrast, the herds affected in the 1953 southern Utah die-off exhibited sores on their faces and discolorations in their wool and did not die immediately. Peterson took several of his animals' carcasses to the agricultural experts at Utah State Agricultural College. He recalls, "They said they didn't know what it was. It looked a lot like, there's a poison plant disease called bighead, and the sheep looked very similar to those that died of bighead, with their heads in a bad condition, swelled up and irritated." Like the southern Utah ranchers, Peterson fell back on commonsense local knowledge to evaluate this possible explanation: "Bighead only came in the spring of the year, and this was in the fall, this was going out, just before the first of December, cause we were right close to that Pilot Mountain range." Experts at USAC confirmed his opinion. Peterson recounted, "They had the poison plant system going up there, probably one of the best in the nation for identifying poisonous plant deaths, and said it definitely wasn't a poison plant but they didn't know what it was. And Dad had his pickup and I had a pickup, and we had two pickups piled as high as we could with sheep pelts."

The suddenness of the deaths and the state of the bodies indicate the Petersons' sheep were likely exposed to nerve gas released from the Dugway Proving Ground, located southwest of the Great Salt Lake and roughly a hundred miles from the Peterson herd. "An expanse of Great Basin high desert the size of Rhode Island, Dugway's sage flats, salt flats, and stony rolling hills were ideal for experimenting with munitions filled with nerve agent," writes local author Chip Ward. "Between 1951 and 1969 there were 1,635 field trials. More than 55,000 chemical rockets, artillery shells, bombs and land mines were blown up to understand how they could be used. Airplanes flew over test grids and sprayed nerve agents to see how it was dispersed in various weather conditions and at various heights." Open-air testing of nerve gas was finally discontinued in 1968, after more than six thousand sheep died downwind in Skull Valley, Utah, following a test of 2,730 pounds of the nerve agent VX in a test code-named Operation Combat Kid. Some livestock owners received a quiet restitution payment from the army, which never admitted fault in the deaths.[50]

The Peterson family was never compensated for the sheep that died overnight in the Hansen Valley. "It was a real loss," Peterson recalled in 2006. "But that wasn't the main loss." Like the Cedar City ranchers, the Petersons incurred their greatest financial hit in the spring, when the ewes began to miscarry their lambs. "We had all those new sheep sheds, well, we should have been getting a hundred and twenty percent lamb crop. We hired the best Basque sheepherders to come and help us run those sheds. We had running water, electricity, everything, really modern out there at the time, probably the most modern sheep sheds in the country," Peterson described. "And we never got seventy percent lamb crop, because the ewes all aborted their lambs. I'm not exaggerating when I say we'd take pickup loads of aborted lambs out to bury. So I had no idea, Dad, nobody knew what it was." The Petersons took several of the lambs to USAC, but its experts had no answers. "They couldn't figure out what was the matter, and they gave us all kind of antibiotics and stuff. And we tried to figure it out, but we never could get exactly what happened." Reflecting on the incident several decades later, when revelations about bomb testing had been well publicized in the region, Peterson "finally kind of figured out that it had to have been radiation." While nerve gas exposure was the more likely culprit, it is impossible to confirm to what type of toxin the rancher, his herders, and his sheep were exposed. No matter what the toxin was, it was almost certainly a product of Cold War weapons development given Peterson's location. Like McRae Bulloch, Peterson drew a parallel between animal and human deaths: "It wasn't only the sheep that died. Every one of the herders that was with that group got cancer and is dead. There was three of them with that herd, and two with another herd, and they're all dead from cancer. It makes you kind of wonder. . . . And I've had cancer too."[51]

Similar problems befell families raising sheep on the Navajo reservation, where ranchers often utilized abandoned uranium sites for watering and sheltering their sheep. In 1988 the Navajo Abandoned Mine Lands Reclamation Department hired Perry Charley to inventory toxic uranium sites on the Navajo reservation that need cleaning up. His inventory located "approximately 500 mines that were used by local shepherds for livestock pens. A well-insulated sheep pen, but potentially dangerous to the sheep that used the mines for shelter and lambing." The ranchers noticed problems in their herds, but with no knowledge of radiation, they had no way to connect their sheep's health issues with the mines or watering holes. A woman "in the Tse-Tah area in northwestern Arizona" told Charley, "I always wondered why my lambs were born without fleece and why some were born deformed or with limbs missing." Charley learned the woman "lived about 200 feet away from a mine for many years, using the mine for a livestock pen. The family was in the process of building an addition to their home using the radioactive mine waste."[52]

The mines that Charley discovered were being used to shelter herds were predominantly on the eastern half of the Navajo reservation. To the west, as the uranium boom waned, the uranium industry usually left behind open-pit mines. Journalist Judy Pasternak described them as posing a different set of dangers: "New lakes had appeared . . . in the arid scrublands. Instead of hunting for puddles in the sandstone, [Navajo sheep-owner Lois Neztosie] could lead her 100 animals to drink their fill." The new lakes that Neztosie and other ranchers increasingly relied on were the abandoned open-pit mines that had filled with snowmelt and spring runoff. The rancher noticed her flock beginning to decline in health. "The birthrate dropped, and the few new lambs that did appear had a hard time walking. Some were born without eyes." Neztosie told Pasternak she also drank from the new lakes, after first "parting the film on the water's surface with her hands." Eventually, "Lois grew to depend on them as she ranged far from home. . . . She got in the habit of filling and refilling a small container with her drinking supply as she moved from one 'lake' to the next, watering her herd."[53]

Other stories are told around the region about sheep lost to radiological pollution from the test site. Gilbert Fraga worked as a radiation monitor for the AEC during the 1950s. He told Carole Gallagher the story of "running into an American Indian that had World War II gunnery medals." Fraga recalled the man "got to crying one time, talking about how badly his sheep were doing on account of beta burns. Tears came down his cheek." Fraga told Gallagher how the man reached into his pocket and revealed "about four medals wrapped up

Fig. 2. Illustration of a cowboy watching a mushroom cloud from the *Atomic Tests in Nevada* pamphlet published by the Atomic Energy Commission (March 1957) and distributed in the downwind region. From the collection of Preston Jay Truman.

in toilet paper in his shirt pocket. He said, 'I was a good Indian and I was a hero then, but now they treat me pretty badly.'" The man and his medals "really got to me. I almost cried," Fraga remembered. "It was confirmed, [the sheep had] beta burns." While it is impossible to know whether the Native sheepherder understood his sheep had beta burns, his encounter with the AEC radiation monitor makes it seem likely that as with the Cedar City ranchers, he had herded his sheep close enough to the test site to make the connection between the government's explosions and his sheep's ailments.[54]

Ultimately, the vast majority of unusual deaths, illnesses, and abnormalities witnessed by livestock owners across the American West during the Cold War's arms race would remain unexplained and uncompensated, even if they were eventually connected to radiation or other forms of weapons testing. That locals in Cedar City correctly alerted government officials to the warning signs of nuclear contamination and were so summarily dismissed only heightened the sense of betrayal and abandonment many downwind residents felt in later years. This betrayal took on an even more sinister tone for those who realized that the contamination stretched beyond their region. Annie Corry remembered, "Carcasses were marketed that were full of radiation. They didn't stop us from selling anything for food."

The phrase "never seen anything like it before," so often invoked in discussions of the sheep deaths, is telling. If that observation is the only evidence that we acknowledge from the ranchers, aside from a few heartrending anecdotes about dead lambs, then we continue to deny them a wealth of knowledge, skill, and agency. Raising sheep is not something one does on a lark while sitting atop a horse, contemplating wide-open western spaces. "Never seen anything like it before" plays neatly into a preconceived notion of rural ignorance. In fact, just the opposite is true. The assertion implies the narrator is drawing upon a cultural system of common sense drawn from local, experiential knowledge of forage conditions, weather patterns, plant characteristics, diagnosis and treatment of a variety of diseases, predator management, breeding, and lambing. While this experiential knowledge did not win the ranchers' case in court, it resonated powerfully with other citizens of the downwind region, providing what was for many unimpeachable evidence that a toxin had entered the local environment, threatening not only their economy but also the integrity of their locally produced food.

{4} LOCALLY GROWN

[W]e plant our garden, but it does not grow big. Years ago, before they did any mining in the mountain, when we planted the plants would get very big, such as watermelons and squash. It is the same in Shiprock; the farm produce are small. The melons are small.

Tommy James, Cove, Arizona

In many corners of twenty-first century America, eating locally grown food has become something of a political statement. Those who choose to do so can take advantage of a proliferation of farmers' markets and local produce sections in grocery stores. An abundance of literature exists to guide the aspiring local eater, from how-to guides to philosophical treatises on the culinary and ecological rewards of rediscovering the local foodshed. There are even T-shirts and bumper stickers for the new dietary demographic, sometimes called locavores. The trend has gained so much popularity that the *New American Oxford Dictionary* selected the term "locavore" as its 2007

Word of the Year, honoring the efforts of the group of San Francisco women who coined it and helped galvanize the movement in 2005. "Our food now travels an average of 1,500 miles before ending up on our plates," Jen Maiser and her cofounders note on the Locavores' website. "Because uncounted costs of this long distance journey (air pollution and global warming, the ecological costs of large scale monoculture, the loss of family farms and local community dollars) are not paid for at the checkout counter, many of us do not think about them at all. . . . The distance from which our food comes represents our separation from the knowledge of how and by whom what we consume is produced, processed, and transported."[1]

The idea of a tomato traveling fifteen hundred miles to be diced into our dinner seems counterintuitive when many of our great-grandmothers grew bushels of them in their backyards. How did we get to this point? Since the late 1800s along with every other form of production, food production has been industrialized. As scientists made sense of the natural world, learning the keys to soil fertility, the properties of disease resistance, and the processes of animal reproduction and growth, industry transformed those insights into commercial innovations. Some of these innovations came from unlikely places; for instance, commercial pesticides had their provenance in chemical warfare research. Pesticides, fertilizers, enhanced animal husbandry, and other new techniques enabled farmers to grow more food, faster, with less labor, and in less space. Philosopher Bernard Rollin notes that agricultural productivity doubled between 1820 and 1920 and "continued to double in much shorter and ever-decreasing time periods" thereafter.[2]

As the Bullochs herded their sheep across Nevada in the decades following the Second World War, the country was experiencing one of the most dramatic increases in agricultural productivity in human history. Transportation networks were advancing as well, and interstate highways, global shipping systems, and improved refrigeration technology made it possible to move food across great distances. New state-of-the-art preservation techniques kept perishable items from spoiling for many months. This highly industrialized system could cheaply market a wide variety of foods to consumers, even if those items were not in season or produced locally; thus shoppers could purchase the Chilean tomato, the Australian beef, the French cheese.[3]

At the dawn of the twentieth century, when one in four Americans lived on a farm, a movement dedicated to eating locally would have been irrelevant. While certainly city dwellers' children might have been unfamiliar with the origin of their milk or eggs, most citizens had some involvement in the production of their food. Even if they did not raise food to sell it, many families kept household gardens and livestock at the turn of the century. The typical farmer would have cultivated "a dozen different species of plants and animals," according to journalist and food activist Michael Pollan. On a farm in Iowa, for example, Pollan notes, "there would have been a fair amount of corn . . . but also fruits and other vegetables, as well as oats, hay, and alfalfa to feed the pigs, cattle, chickens, and horses." While a portion of those crops would have likely been sold and shipped elsewhere, the family and others in their community would have consumed much of this produce.[4]

As the century progressed, mechanization and chemical and technological advances reduced the amount of labor needed to oversee agriculture, leading to bigger operations run by fewer people. "Just before World War II, twenty-four percent of the U.S. population was involved in production agriculture; today the figure is well under two percent," Rollin explains. Today more Americans are in prison (2.3 million) than in farming (less than 2 million), and many of those farmers who remain cannot actually feed themselves or their families with the food they produce. Their crops, predominantly corn and soybeans, are "commodities that must be processed or fed to livestock before they can feed people."[5]

During the atmospheric testing era and the uranium boom of the 1950s, many small towns had not yet been transformed by industrialized agriculture. In the communities of the downwind region, eating locally produced food was still a way of life rather than a lifestyle choice. Staples such as milk, meat, fruit, grains, and vegetables were still produced and often consumed in the communities that grew them. By keeping gardens, preserving food for the winter, and buying from local farmers, individuals were maintaining a culture of food production they had learned from their parents and grandparents while participating in a community-based economy. While not every individual farmed or ranched, it was difficult to remain ignorant of the path food traveled to reach one's plate and of the labors others had undertaken to produce it.

Mary Dickson, who grew up in Salt Lake City in the 1950s, remembered, "When we were kids, people all had gardens. I mean everybody grew food and you bottled your own pickles. . . . My grandfather up in Morgan did, and we would go up there all the time and eat stuff we picked out of the garden. And he would take us to nearby farms and we would drink milk straight from the cows. They'd grow all their own food." She remembered visiting them and watching her grandmother preserve food. "You would bring home jars of stuff."[6]

With that local knowledge—and that homegrown tomato or jar of neighborhood fruit preserves—came a sense of pride, tradition, and shared ownership. Ecological awareness was still a nascent phenomenon in popular culture in the late 1960s and early 1970s, yet individuals in rural communities were already participating in a generations-old system of knowledge that recognized the interconnectedness of the atmosphere, watershed, flora, and fauna even if at times that knowledge was imperfect or abused. They understood, generally, that contaminating any one of those elements could permeate the other components of their surroundings and eventually reach their own bodies, because they saw much of the process firsthand.

The introduction of radiological pollution into the downwind foodshed was rarely illustrated as dramatically as it was in the sheep

☰ OUR SOIL ★ OUR STRENGTH ☰

Fig. 3. "Our Soil, Our Strength" graphic from *Iron County (Cedar City UT) Record*, 1 November 1956.

die-off in Cedar City, Utah, yet those who raised their own food were accustomed to paying close attention to environmental conditions and often recognized subtler indications. They watched when streams carried off uranium mine waste during flood season, when the wind picked dust up from uranium tailings piles and scattered it onto their gardens, when fine material drifted onto backyard vegetable plots and alfalfa fields from strangely colored and hazy skies after massive flashes or booms occurred to the west. They paid close attention when animals and crops failed to thrive and actively sought to explain such failures.

Kay Millet, who raised her family near Cedar City during the atmospheric-testing era, remembered one winter when she started tomato plants indoors, intending to plant them later in the spring. "It was February, a nice sunny day, and I thought, 'I'm just going to set these tomato plants out and let them get sun.'" A short time later, Millet saw the plants had turned "white, crusty-like and laid over, gone for no reason. That same year when we raised the garden we noticed the squash and tomatoes would get this light stuff on them, the leaves would get white and crusty and the squash would be all yucky and the tomatoes did the same thing." Millet observed the white crust "kind of spread around the garden. We ate them, the ones that were good."[7]

After a few years, most gardeners would be hard-pressed to remember when a specific blight swept through their gardens, but decades later, Kay Millet still remembered when it happened. "That was 1957, or '58, because Sherry [Millet's daughter who succumbed to leukemia at age five] was just a baby then. They were doing tests all the time." Millet lived close enough to the test site to know that around the same time she witnessed the strange phenomenon in her garden, nuclear testing was occurring regularly to the west. For those who did not live in the immediate vicinity of either a test site or a uranium-mining facility, other clues would be necessary to connect crop and livestock problems to radiological exposure.

Born in 1948, in northern Utah's Duchesne County, Dave Timothy grew up in the dairy industry. He spent his childhood and teenage years working on his father's dairy farm in Altonah at the base of the Uinta Mountains, a steep, massive range east of Salt Lake City. From the age of eight or so, Timothy "milked cows . . . irrigated, plowed, harrowed, disked, leveled, planted, cut the hay, baled it, raked it, hauled it, fed the calves, weeded the garden, [and] mowed the lawn. . . ." His father's "was a Grade A dairy. This was a first-class dairy. [Milk] was taken through pipelines, glass lines, to the tank, where it was immediately cooled, kept at a constant temperature until it was picked up by the milk truck." Timothy's father sold his milk to Hiland Dairy, which in turn distributed it across the region.[8]

After they had finished milking the cows each day, Dave and his father, James, took home two or three gallons for their family's consumption. Between Dave, his two younger sisters, and their parents, "[t]here was very seldom excess milk" left over the following day. "Milk was our produce that we had. So, many of our meals, the basic . . . diet was around things we raised. We raised our own vegetables from our garden and then milk being another basic. Things were planned around and developed from that." The Timothy family also consumed local beef: "A lot of times . . . [father James Timothy] would get a beef [cow] from one of our neighbors that had fattened it, grain fed, finished it out. We would sell ours or trade for one that was more tasty." Occasionally, Dave recalled, "if you were lucky that year," a hunting trip brought home local deer meat. His mother, LaWanna, "can[ned] a lot of the garden produce so that we would have it through the winter," and the family bought "apricots, apples [and] peaches" from neighbors. Drinking water came from a spring on the property, and crops and gardens were irrigated with snowmelt runoff from the Uinta Mountains watershed.[9]

Ivan Sidney, who was born the year before Dave Timothy, grew up on a farm roughly four hundred miles due south. Sidney lived with his mother and his grandparents at First Mesa, near the town of Polacca, on the Hopi reservation in northern Arizona. When Ivan was five years old, his father, a World War II veteran, had died following an on-the-job accident while working on a Bureau of Indian Affairs road construction crew. Sidney's mother was plagued with rheumatism, so her eldest son Ivan "grew up being her hands." He described a childhood "speaking only Hopi. . . . I grew [up] pretty much in the traditional way." The family lived without electricity, "no phones, and no refrigerators, and we drank mostly from the spring. I remember going with my grandfather to get water. When we wanted the coolest water in the hot summer [we would] go to the spring right at lunch and get some water . . . and much of our food was from the farm, corn, watermelon, beans." In addition to the produce they cultivated, Sidney's grandparents also raised livestock to provide the family with meat. "I was raised on sheep, a lot of us did our own butchering. We don't just eat the meaty part of it—lamb chops and all—our traditional foods even included intestines. Nothing went to waste. There was something called bloodcakes, too, from the blood. What an animal ate, we'd take it too." Sidney labored alongside his grandparents on the farm, hauling water, "clearing sandstones, . . . pulling weeds, things like that."[10]

As children growing up in remote farm communities several hundred miles from the Nevada Test Site, Dave Timothy and Ivan Sidney had no knowledge of the bomb tests as they occurred. Years later, Timothy remembered a number of times "working in the fields . . . when it would be very hazy, just super hazy. It was almost like it had been a real bad windstorm, but there hadn't been. I remember commenting to Dad quite often about different times that it sure was strange to see it that hazy." Timothy remembered the haze being "almost like a smoky color, but it wasn't smoke." On the Hopi reservation, Sidney recalled, "we're so isolated, we didn't know what was going on" at the Nevada Test Site. Several Hopi had battery-operated radios, but "reception was very poor out here; the only reception we had was at night, from Oklahoma. They must have had a powerful transmitter, because I remember it was called KOMA. Everywhere in this village you could hear it just echoing. They played the old-style music. . . . If there was any kind of warning" about nuclear danger, Sidney declared, "we wouldn't [have] hear[d] it."[11]

Of course, few warnings were broadcast, and certainly none from radio stations outside the immediate downwind region. The Atomic Energy Commission occasionally announced a test ahead of time so citizens could view it from Las Vegas or from the rooftops of southern Utah towns, but these messages included no indication of danger beyond cautioning viewers to wear sunglasses to protect their eyes from the brightness of the blast. Families such as the Timothys and the Sidneys would never have heard these broadcasts, as the AEC saw no reason to inform citizens outside the immediate vicinity of the test site. The commission had its hands full simply managing the information it provided to those people who lived close enough to witness the tests.

After the Upshot-Knothole test series of 1953, patriotic press releases were no longer adequate to keep citizens in the immediate vicinity complicit. One test in particular that year bore especially dramatic results. Shot Harry was hardly the largest weapon detonated in Nevada, but it produced the single-greatest quantity of external gamma ray exposure of any domestic nuclear test. Alarmed the post-blast cloud would cause serious radiation sickness in citizens immediately downwind, the AEC did issue a rare radio warning in St. George, Utah. Later it was reenacted in the 1955 AEC film *Atomic Tests in Nevada* (the same film that erroneously claimed livestock owners received warnings prior to nuclear tests). Filmed in Technicolor and featuring a cast of St. George residents, the film sought to quell the fears of residents in the immediate vicinity of the test site.[12]

It opens with scenes of the early morning streets of St. George. Except for the policeman, the gas station attendant, and the local milk-man, "everyone [was] asleep," the narrator intones. "Only our night owls saw it, that great flash on the western sky, an atomic bomb, at the Nevada Test Site, 140 miles to the West. But it is old stuff to St. George. Routine. They've seen a lot of them, ever since 1951. Nothing to get excited about anymore." As the morning progresses, the "thriving community" goes about its business. Women are shown hanging their wash, peeling potatoes, and sending their children off to school while men fill their cars at the service station and patronize local businesses. When the AEC's radio announcement informs them "that due to a change in wind direction, the residue from this morning's atomic detonation is drifting in the direction of St. George," residents calmly vacate the streets, taking cover indoors until their radios notify them the danger had passed.

On the actual day of Shot Harry, few downwind residents actually heard the warning. Many were already at work, away from their radios, and others had no radio at all. Quentin Nisson, mayor of the nearby community of Washington City, Utah, from 1950 to 1964 and proprietor of the Washington Mercantile, received the warning via telephone, as Washington City was still four years away from receiving radio service. The AEC placed a phone call to Nisson at the Mercantile around "10:00 or 10:30" a.m. and asked him to go to the elementary school and order the children to be kept inside during recess. Nisson walked across the street to pass the warning along to the school. Luckily, he recalled, "recess wasn't till about 11:00, you know, so they did keep them in during recess here. . . . But I remember when I went over [to the school] seeing that there was kind of this yellow-looking stuff . . . [the fallout] was already here."[13]

Even in St. George, the warning did not reach everyone. Frank Butrico, St. George's AEC radiation monitor, recalled, "It wasn't too much of a surprise that not everybody had the word. Cars were still on the road within St. George, people were still walking on the streets, and most distressing, when we passed the grade school, we noticed that the children were still on their morning recess, the teacher having not received the information about taking cover." Butrico received instructions from the AEC to "be sure and discard my clothing, and to be sure and keep showering until I reduced the amount of radiation that was on my body." He asked his superiors "whether we should be doing the same thing in an announcement to the people in the community, and of course the answer was a resounding no, because this would create a panic situation."[14]

The AEC screened *Atomic Tests in Nevada* for the St. George Chamber of Commerce in April 1955. Member Sheldon Johnson recalled chamber members felt a "kind of pride" at this depiction of their participation in the work of national security. "We were recognized as somebody. . . . And up till [then], St. George wasn't anything." The film subsequently played for local residents, who found it exciting to see their streets and neighbors on-screen. Years later, St. George's Technicolor debut gained darker significance in downwinders' memories. Hardware store proprietor Elmer Pickett, who was shown in the documentary listening to the radio in his store, noted that many of the other film participants later succumbed to cancer. "Remember the milkman and the police? They had a housewife and a sheriff . . . out of the bunch that was in that film, two-thirds of them died with cancer."[15]

The film had been produced as part of the AEC's new campaign of "zone management," which sought, in Frederick Schoemehl's words, to "reshape the political and cultural geographies of the region." Subsequent to the sheep die-off and the call for citizens to go indoors after the Harry test, the AEC "declared 'complete acceptance' of atomic testing as its goal. . . . A process of public 'indoctrination' began with a view that the off-site region was a network of zones that the AEC would manage." Within two years of Shot Harry, test site officials had become "obsessed with local matters, a recognition that on- and off-site activities were, in reality, inseparable." The zone management campaign institutionalized the sort of condescension and disrespect AEC officials involved in the sheep death investigation directed at the ranchers. Schoemehl argues the zone management program "embraced an institutional arrogance and authority. Common people—the ranchers, miners, and rural residents who occupied the region—remained subjects of management and manipulation." A key technique of the zone management program was the AEC's wooing of local leaders, who could then "carry [the AEC's] message to the bulk of the local population," a strategy that Schoemehl observes was "not dissimilar from colonial systems of governance."[16]

Flattered by their participation in the great work of national security, as depicted by zealous AEC officials and *Atomic Tests in Nevada*, many local leaders (though not all) enthusiastically endorsed the AEC's messages and placated worried citizens. The AEC recruited radiation monitors from the Public Health Service and assigned them to various communities and regions downwind of the test site, where they distributed badges to measure radiation, shared pamphlets and films with locals, and "educated" physicians about radiation-related health problems. The 3 March 1955 *Washington County (St. George UT) News* advised concerned citizens that AEC representatives were stationed locally, and readers could easily contact them with questions by dialing "724, at the Rugged West," a motel located in the center of downtown St. George, across from the popular Big Hand Cafe and the new J. C. Penney store.[17]

The AEC set out to "manage" downwind perceptions of the test site in Tonopah, Nevada; St. George; and other towns because their residents were already well aware of the ongoing testing. The townspeople regularly heard the sound of the blasts, felt the shock waves, and saw the bright flashes of light and the distinctively shaped mushroom clouds. By the time these clouds passed over distant communities such as Altonah and Polacca, they had dispersed enough to resemble ordinary clouds and gave no outward indication of their toxic content other than their occasional strange coloring. At that distance, no one saw or heard the initial blasts. If a radiation monitor did cross paths with a citizen in these communities while measuring for radioactivity, he would have little need to explain himself. Regional newspapers such as Salt Lake City's *Deseret News* did discuss test site activities periodically, but people outside of the immediate downwind area had little knowledge of bomb detonations, precisely as the AEC intended.

Unaware of the explosions to the west, the Timothys and the Sidneys and other families went on with their lives. Their parents and grandparents continued to run their farms, and both young men set out into the world. Ivan Sidney went off to a BIA boarding school in Phoenix for high school, and he allayed his homesickness by learning guitar, a passion that would sustain him through many difficult periods to come. After graduating, he enrolled at the newly formed U.S. Indian Police Academy in Roswell, New Mexico, and several years later was chosen as the chief of police for the Hopi Tribe. In 1981, Sidney became the youngest person ever elected as the tribe's chairman.

In 1996 Sidney began to experience severe flu-like symptoms while working as the tribal liaison for the president of Northern Arizona University. "I went to the local Indian Health [Service (IHS)], and they would diagnose it as the flu, and give me medication for that. And I'd be okay for a little while, but it became more frequent." Sidney's job at the university had enabled him to acquire health insurance, which he credits with saving his life, by offering him access to cutting-edge diagnosis and treatment technology off the reservation. "If I didn't have my own insurance, and being just a ward of the government, having services available from IHS, I'd be dead right now. And that's how I'm afraid a lot of our Indian people are dead today because they don't have any insurance." Diagnosed with non-Hodgkins lymphoma, Sidney underwent eighteen months of chemotherapy and a blood stem cell transplant. He attributed much of his recovery to his wife, who prepared him "mostly Hopi foods, foods that my mother and grandmother cooked. They were just common things." Thus nourished, Sidney "started comin' through it." It was not until he began to recover that Sidney asked his doctor what might have caused his disease. "His first question he asked me, 'Let me ask you, have you ever been exposed to radiation?' And I said, 'All the x-rays you gave me.' And he said, 'Noooo. Do you remember anything?' Then he said, 'I want to have the nurse give you some material.' That's when I heard about this compensation."

The materials the nurse provided informed Sidney that his family's home lay downwind of the Nevada Test Site and in a region so heavily contaminated that Sidney could qualify for a compensation payment of $50,000 from the federal government if he could legally demonstrate he had resided there during the atmospheric-testing era. Reeling, Sidney began to piece together how his body might have absorbed that contamination. "I drank from local springs. This was our primary source of water. We drank that, not knowing that there was some of the fallout." He considered that the vast majority of the food he and his family consumed was grown locally, in an area he now knew had been heavily contaminated.

Sidney then began to tally those in his immediate family who had also been stricken by cancer: "[M]y mother died from breast cancer, she refused to take chemo. She went on chemo after her breast was removed, but after a short while, she refused to get more, and she just died from cancer." Subsequent to his own cancer battle, Sidney reflected, "I know why she didn't want to take it. I used to say, why did she leave us? If she took chemo maybe we would have enjoyed our mother a little longer, because that [she] was the only thing I had. I forgive her now, because, boy, having chemo, it hurts. And she died." In addition, his mother's "younger sister, who is dead now, also had cancer, but she lived longer. Her other sister, my aunt, just passed away this past August from cancer. My uncle died from prostate cancer. And I, being too busy, I didn't realize I was part of it."

Dave Timothy discovered he too was part of a pattern of strange health problems, only much earlier in life than Ivan Sidney did. In 1967, at age nineteen, Timothy left his parent's farm in Altonah and moved to St. George to attend Dixie State College. Partway through his first quarter, Timothy noticed himself becoming "extremely hyper" and simultaneously very tired. After discovering a lump in his throat, he went to the college library, researched his symptoms, then took himself to the doctor. Shortly thereafter, diagnosed with advanced thyroid cancer, Timothy underwent a radical thyroidectomy on 17 December and radical dissections on the left and right sides of his neck in January. A series of cobalt treatments—a form of radiation therapy—followed. His operations left him without the "muscles in the front of [his] neck that it takes to keep the vertebrae lined up properly," diminished much of the feeling in his shoulders and head, and limited his lifting ability. He continued to battle cancer for many years.[18]

Other members of Timothy's family from northern Utah also struggled with health problems. Among others, his mother discovered frequent lumps under her skin, his sister experienced symptoms of a damaged thyroid, and his aunt June Carrell had her thyroid removed. He recalled, "Many [blood relatives] died from cancer . . . they were most of them farmers at that particular time in that area. They most of them lived out in the Uintah Basin. . . . Uncle Charlie [Timothy] died . . . I believe he had stomach cancer. Vaughn Timothy, . . . Presley Timothy died of leukemia, . . . Pink Timothy . . . prostate cancer, and cancer of the liver. . . . My Aunt Mary, Mary Murray . . . she had lupus, erythematosis, blood cancer."[19]

In the years following Dave Timothy's first bout with cancer, the health problems and fallout exposure of southern Utahns began to be written about in the Salt Lake City newspapers. Naturally inquisitive and still struggling with his health, Timothy sorted through his memories. He recalled the hazy skies he saw over his family's farm as a boy and remembered that around the same time, between 1957 and 1962, "there was a lot of talk about our milk possibly being too hot to sell . . . if the milk was too hot, that's the way it was referred to, if it was too hot that it might have to be dumped. And people were saying, 'If we should have to dump this milk . . . who is going to buy it?'" He remembered school bus conversations with neighbor Susan Fisher, who told him that men were visiting her father's farm and "taking fish from the ponds and that they were taking samples of deer and calves." When he contacted Fisher as an adult, she confirmed his memories and told him that she and other residents of the farms underwent scientific tests in Salt Lake. These tests were intended to measure the presence of radiation in the farm residents' bodies, although many of the subjects probably did not know the full significance of the tests.[20]

Returning home to Altonah, Dave Timothy sought out Carl Carrell, George Fisher, and Ross Munson, dairy farmers of his father's generation who operated their dairy herds within a few miles of the Timothy farm. They told him Dr. Robert Pendleton, a scientist from the University of Utah, had monitored their milk, their farms, and some of their family members, including Susan, George Fisher's daughter, and it was Pendleton who had warned them their milk could be "hot." Hoping for answers, Timothy tracked down radiation ecologist Pendleton at the university. In a 2005 interview, Timothy recalled that when he told Pendleton he had grown up on a northern Utah dairy farm and revealed why he had come, Pendleton, "a big strapping man, started to cry."[21]

Prior to joining the university's faculty, Pendleton had spent most of the 1950s studying environmental contamination and radiation measurement in the employ of the U.S. Army Chemical Corps and Hanford Laboratory in Washington State. During that time, he became interested in the process by which the radioactive element Cesium-137 (Cs-137) became integrated into plant systems. In 1961, after moving to Utah, Pendleton took his research a step further, setting up a network of Utah dairy farmers in order to study the uptake of Cs-137 into the milk supply after dairy cows consumed plants bearing the element. Prompted by scientific curiosity rather than a perceived overt threat to human health, Pendleton's cesium study did not raise any alarms initially.[22]

On 7 July 1962, the radiation ecologist took a group of University of Utah students into Big Cottonwood Canyon, twenty miles southeast of Salt Lake City, seeking to "measure the small quantities of radioactivity that are associated with the disintegration of granites, [and] various kinds of rocks." Pendleton was aware at the time that nuclear testing was under way in Nevada, and when the group's instruments "went completely nuts," indicating the presence of far more radioactivity than disintegrating granites could produce, he surmised they were "in the fallout track" of a recent test. Upon returning to the university, Pendleton noticed a "big rather reddish brown colored cloud" hovering over the mountains to the south. "The following morning," he remembered, "we could find [radioactivity] all over—we counted on our lawns out here." Shot Sedan, a 110-kiloton bomb, had been detonated in Nevada the previous day as part of the Operation Plowshare test series, dedicated to exploring the use of atomic bombs for nonmilitary purposes such as excavating for mines and dams. Maps assembled later showed the trajectory of the fallout cloud passed directly over northern Utah, then arced over the Midwest, and eventually deposited heavy fallout in Iowa.[23]

Unnerved by the high radiation levels present in the Salt Lake City area, Pendleton collected milk samples from several of the farms in his cesium study. Tests revealed the milk to be "highly contaminated" by the radioactive element iodine-131, which posed a significant risk to human health. On 16 July, Pendleton "notified the Utah State Department of Health of this situation and suggested that highly contaminated milk be diverted to the production of cheese, powdered milk, or condensed milk to reduce unnecessary exposure to the population." I-131 has a short half-life of only eight days, and it was hoped that diverting the milk would enable the element to decay beyond a point of extreme danger. There is no evidence that such a diversion occurred.[24]

Pendleton did receive funding to "delineate the extent and degree of contamination by I-131." Using the network of dairy farms he'd set up for his cesium study, Pendleton and several other researchers from University of Utah began to study I-131 accumulation. They found dangerous levels of I-131 throughout the milk supply in northern Utah. The data they gathered led to a startling discovery: The degree of radiation exposure received by populations downwind hinged more on local factors, such as elevation, moisture, and livestock feeding practices, than on the size or characteristics of the bomb or even on the proximity of the downwind population to the test site. Such local variations "caused differences in [the amount of radiation intake] ranging from 2 to 450 fold." The scientists concluded, "It is obvious that evaluation of the hazards from local fallout should be made on the basis of local, intensive monitoring of milk and people, rather than on fission yield, aerosol trajectories, or estimates based on air monitoring." Their research found the most severe contamination "in milk and people from farms where cattle were fed freshly-cut green alfalfa or were grazing on wet meadows. . . . Highest I-131 levels occurred in farms at high altitudes."[25]

Anyone who has lived at the base of a mountain range has seen the rain shadow effect in action. Clouds traveling in the lower atmosphere are pushed upward when they drift against the slopes of mountains, causing their temperature to drop and airborne moisture to condense into precipitation. Since the AEC only conducted nuclear tests when the wind was forecasted to blow east, fallout-bearing clouds were repeatedly swept into the Wasatch and Uintah Mountains, massive ranges in northeastern Utah that butt up against the Rockies to form the eastern wall of the Great Basin. The Wasatch Range, which runs north–south, averages ten thousand feet in elevation. The Uintah Range, running west–east, tops out between eleven thousand and thirteen thousand feet. Trapped by these high peaks, clouds from the test site were transformed into snow and rain, which bore their radioactive content to earth, contaminating watersheds. Farmers in the basins below funneled the watersheds into well-developed irrigation systems that the early LDS settlers had designed and maintained thereafter to capture as much water for agriculture as possible.[26]

"Radioactivity was not raining down on a wilderness," Frederick Schoemehl writes. "It was entering a landscape previously transformed by a century of human reengineering." The problem was more complex than irrigation systems funneling radioactive rain into the food supply. Since I-131 decays relatively rapidly, some of its toxicity likely was lost by the time it reached crops and livestock via irrigation systems. However, even if the I-131 in irrigation water had decayed beyond the point of danger, those irrigation systems still distributed other longer-lasting radioisotopes. Most important, irrigation created a presence of moisture in cultivated fields, primarily alfalfa, that Pendleton and his colleagues found magnified the uptake of airborne I-131 into crops. In northern Utah, farmers living near the mountain watersheds tended to irrigate their fields far more heavily than their southern Utah counterparts did. Cattle fed fresh-cut, moist alfalfa, known as green crop, consumed significant amounts of the isotope, which transferred into their milk. Thus began the process known as bioaccumulation.[27]

"Bioaccumulation means that a blade of grass may contain a tiny amount of dioxin, or some herbicide or pesticide chemical," writer Chip Ward explains. "A cow, however, can eat a lot of grass. So the cow more or less gathers the dioxin or pesticide that is spread out in the pasture's grass and concentrates it in her milk, especially in the fatty cream. A dairy producer then collects the cream from many dioxin-gathering cows, and further concentrates it" immediately prior to human consumption. Ward notes, "Being at the top of the food chain has its price."[28]

Irrigation systems were not the only piece of agricultural reengineering funneling radiation toward the food supply; the alfalfa crop was also to blame. Confronted with a largely arid climate, early LDS settlers had sought out a crop that could produce well in dry conditions to augment the native plants, which could support only so many dairy cattle. "Alfalfa supplied the link," Schoemehl notes, "provid[ing] a way to feed ever-larger herds, thereby circumventing limits that otherwise would have been imposed by native plants." With alfalfa crops spread out over the countryside, farmers had sizable herds of dairy cattle, which consumed prodigious quantities of fallout material and thereby compromised a large regional dairy industry.[29]

The highest levels of I-131 contamination discovered in Pendleton's research were not in milk from St. George or Cedar City, close to the test site. They were found in the milk supplies of Cache County, bordering Idaho. Duchesne County, home to the Timothy farm, was not far behind. Pendleton stated a "considerable amount of radioactive material came down in the water from the Uintah [Mountains]." Owing to the proximity of the dairy industry to most of their communities, residents of the Great Basin region tended to consume milk soon after it was procured from the animals. Consequently, children consumed the contaminated milk while it still bore high levels of I-131, which traveled directly past their developing thyroid glands as they swallowed. Milk was not the only contaminated food source. Pendleton's samples of local produce, beef, wild game, and water collected in the region in the 1960s revealed that radioactive byproducts of nuclear testing had contaminated all avenues of the Timothy family's food supply.[30]

Pendleton's attempts to warn the public via the Public Health Service and other government agencies failed. Testifying on behalf of Dave Timothy, who sued the federal government for damages in 1981, Pendleton recalled, "The argument was, 'Don't say anything of this kind. It's going to panic all the people, and they will do something that is going to be hard for them to live with later. They might drink too little milk and depress their calcium intake.'" Concerned over the public health risk that I-131 posed, particularly to children, Pendleton and his wife "spent a couple of nights driving all over most of northern Utah going to station after station after station . . . talk[ing] them into taking their people off radioactive milk and using powdered milk or canned milk." By "station" Pendleton meant the dairy farms where his monitoring equipment was located. When Dave Timothy found him in his office years later, Pendleton told him he had suffered nightmares about the fate of the children from the dairy farms he monitored as children were being urged to drink more cow's milk than ever before.[31]

During the atmospheric-testing era of the 1950s, dairy products came to symbolize Utah's agricultural bounty, corresponding with a nationwide upsurge in dairy consumption and marketing. In 1946, President Truman signed into law the National School Lunch Act, "as a measure of national security, to safeguard the health and well-being of the Nation's children and to encourage the domestic consumption of nutritious agricultural commodities and other food." The House Committee on Agriculture intoned, "Not only is the child taught what a good diet consists of, but his parents and family likewise are indirectly instructed." The central tenet of that dietary education was to encourage regular dairy consumption; indeed the act dictated that children's lunches were to include a half pint to two pints of whole milk every day. It also stipulated that federal money would not flow to schools that failed to include milk in the required amounts: "Reimbursement rates for lunches served without milk were reduced by two cents, but this was permitted only if an adequate supply of milk meeting state and local standards as to butterfat and sanitation was not available; otherwise, meals without milk were not reimbursable."

Nutritionists did not develop these standards alone. Journalist Barry Yeoman points out the program was "designed to subsidize agri-business, shoring up demand for beef and milk." The program has changed little in the decades since it was launched. Yeoman quotes nutritionist Jennifer Raymond, who sums up the school lunch program as "a welfare program for suppliers of commodities. It's a price support program for agricultural producers, and the schools are simply a way to get rid of the items that have been purchased." It was a sign of the times that in the School Lunch Act, Truman linked nutrition with national security. Well-fed, strong Americans were often invoked as a defense against communism, and purchasing American commodities was considered just as crucial, if not more so.[32]

Milk producers were more than ready to hop on the dairy promotion bandwagon, and they appreciated the parallel being drawn between milk consumption and patriotism as it was good for business. Midway through the 1953 Upshot-Knothole nuclear test series, and around the same time of the catastrophic Cedar City sheep die-off, the newly formed Utah Milk Foundation initiated an extensive advertising campaign to boost milk and dairy product consumption. Playing on local pride and patriotism, the foundation claimed "membership drawn from nearly all of the state's 2,500 grade A milk producers." Not only did milk taste "extra good in Utah," consumers were reminded, but "America would be an even stronger nation in terms of economic and human health if Americans consumed more milk." Foundation president and dairy farmer Eugene Pace reminded consumers that the GIs serving overseas in Korea that year "voted milk as their favorite food." The campaign not-so-subtly suggested that drinking milk not only would strengthen consumers' bodies but also would fortify the health of the nation, which was embroiled in fear over perceived threats from communist enemies.[33]

The foundation designated June 1953 as Dairy Month and kicked off events with the coronation of "a milk-drinking queen—Utah's Dairy Darling." University of Utah co-ed Lu Ann Richards professed

DRINK MILK

...tastes extra good in Utah!

UTAH MILK FOUNDATION

Fig. 4. "Drink Milk . . . tastes extra good in Utah," Utah Milk Foundation advertisement from *Iron County (Cedar City UT) Record*, 12 May 1955.

to drink "a quart of milk a day, and 'love[d] nippy Swiss cheese and ice cream with nuts in it.'" The Safeway grocery store chain offered a Studebaker Champion four-door sedan as the grand prize in a state-wide essay contest on "the reasons why Dairy Month deserves the support of every Utah citizen." Local dairies sponsored milk carton drives in their communities' schools. In Cedar City four-year-old Marsha Lewis gathered 6,074 milk cartons to win a bicycle from Arden-Sunfreeze Creamery.[34]

Milk was such an integral part of daily life in the 1950s that fifty years later baby boomers in the downwind region could still recall where their family's milk came from. Cedar City resident Claudia Peterson's milk came from "neighbors down the street." Gwendolyn Nisson of Washington City fed her children milk from her father's dairy farm. Even people who grew up in urban areas can still name the dairy that bottled their milk. In Salt Lake, Janet Seegmiller's family bought milk from a Cache Valley dairy, while Mary Dickson drank the milk from Salt Lake's Winder Dairy, which, along with many local dairies, delivered fresh milk to customer's homes every morning. Michelle Thomas remembered drinking raw milk from the dairy of St. George farmer Rulon "Boots" Cox until her mother became concerned by rumors of deformed calves being born to local dairy herds. She switched her family to milk from northern Utah's Hiland Dairy—the very dairy the Timothy family in Altonah was supplying—hoping it might be safer.[35]

Government agencies responsible for public health had indications that radioactive contaminants had entered the food supply, making the fervent promotion of milk drinking all the more unnerving. In addition to measuring local radiation levels and noting local sentiments about testing, the AEC's Public Health Service monitors also gathered milk samples from southern Utah dairies all while using "extreme care . . . and diplomacy in order to avoid arousing unwarranted doubts in the minds of the dairy farmers concerned." St. George dairy farmer Boots Cox supplied samples of milk, feed, manure, and the occasional cow to the AEC. Years later, he told Carole Gallagher, "I didn't know what it meant, or nothing. . . . I asked them a lot of times, and they told me that they would tell me if they ever found any traces of radiation. The only time they ever did tell me was when China let off their first explosion, and they told me then that they got a slight trace." Despite the economic blow it would have dealt him, Cox told Gallagher he would have been willing to dump contaminated milk had he been told it was dangerous. "I asked quite a few times, but they always said no. I didn't know enough about it to know whether to suspect anything. . . . Knowing what I know now and read and everything, I don't see how they could have helped but know that there was fallout in it."[36]

While the data that zone monitors gathered on milk contamination in the 1950s was never made public, Frederick Schoemehl argues the installation of a federal milk-monitoring system demonstrates the AEC's knowledge of the incursion of radioactivity into the food supply. It is difficult to draw conclusions about the complexity of that knowledge. Scientists were still learning to identify both all of the radioactive by-products of atomic detonations and their associated effects. In the 1950s, strontium-90 was considered the primary threat to human health, largely because of the efforts of chemist Willard Libby, who was considered a leading authority on fallout. Formerly with the Manhattan Project, Libby was appointed to the AEC in 1954 and oversaw the commission's fallout-monitoring program, reassuringly named Project Sunshine. Strontium-90 was the primary research focus of the program, which took contamination of the food chain into account but failed to reflect completely or accurately the extent of that contamination. At a 1957 congressional hearing on radiation safety, AEC health physicist Forrest Western suggested the dairy cow was "actually a protection," the theory being the body of the dairy cow "shielded" the public from danger by filtering out strontium-90 before it entered the milk supply.[37]

The AEC's admission that strontium-90 had entered the food chain did motivate inquiries into the safety of the food supply. The March 1959 edition of *Consumer Reports* offered readers "a study—the most thorough of its type yet undertaken—of the effects of fallout on 'The Milk We Drink' [b]ased on laboratory tests of samples collected from 50 cities across the United States and Canada." The report provided rudimentary information on radiation filtering into the food supply and the body and concluded that "there is incontrovertible evidence that the strontium-90 content of milk has been increasing since 1954." The report ended by admitting its authors had no "clear recommendation. None exists. . . . We can surmise that we still are not heavily dosed, but we can also be sure that there have been unattributed individual tragedies caused to persons by fallout."[38]

Public reporting on the incursion of radioactivity into the food supply helped bring about an agreement between the United States and the Soviet Union to enact a temporary moratorium on nuclear testing. Despite the moratorium, the news about milk contamination spurred citizens to action. Horrified that a staple they fed their children might contain radioactive poisons, mothers across the country joined antinuclear organizing efforts. To protest the resumption of testing in 1961, "on November 1, seemingly out of nowhere, an estimated 50,000 women in more than sixty cities walked out of their kitchens in a one-day strike." Organizers of the strike announced the action "through female networks: the PTA [Parent Teacher Association], the League of Women Voters, the Women's International League for Peace and Freedom. They even used Christmas card lists," journalist Ruth Rosen wrote. "After a decade of containment and the Cold War, with citizen dissent silenced . . . Women Strike for Peace [WSP] stunned the nation." At the heart of their campaign was the threat that the nuclear buildup posed to their children. The WSP "activists carried placards demanding such modest goals as 'Pure Milk, Not Poison,' and 'Let the Children Grow.'" On 12 July 1962 WSP protesters picketed with those messages at the AEC office in Las Vegas. Three days later, forty-four WSP women picketed at the entrance to the test site itself.[39]

The *Consumer Reports'* study and WSP's pickets seem to have provoked little corresponding outcry in the immediate downwind region, likely because the AEC's zone management program was effective and residents adopted the LDS church's unfavorable attitude toward public protest. If downwind residents did read the *Consumer Reports'* study, they might have actually been reassured: the figures offered for the strontium-90 content of milk produced in the Intermountain West were some of the lowest in the nation. For the purposes of comparison, the strontium-90 content of milk sampled in Salt Lake City was half that of Seattle's and one-third of that found in New Orleans. The presence of strontium-90 in the milk supplies of cities distant from the Nevada Test Site illustrates just how little control the AEC had over the dispersal of radiation. Wind currents regularly carried debris from the Nevada tests beyond the immediate downwind region, and periodic changes in wind direction made predicting where that debris would end up difficult. Nuclear tests conducted elsewhere in the world at that time—for example, in Australia, the Pacific, Kazakhstan—also contributed to the presence of radioactivity in global air currents and international food supplies.[40]

Those residents in the downwind area who did discern contamination in their food supply did so largely on their own time and via their own observations. There were other indications about the presence of radioactivity aside from the AEC's monitoring or Pendleton's warnings. When Dave Timothy returned to Duchesne County in the late 1970s to talk to the dairy farmers, he also spoke with former uranium prospectors Jimmy Bird, Don Birch, Max Birch, Junior Hicks, and William Durfey. They related how, in the early 1950s, many of them filed mining claims, "thinking they had vast deposits of uranium," only to find out their Geiger counters had picked up "background radiation . . . because of 'the bomb,' as they referred to it." Government officials in charge of authenticating uranium finds for the AEC were the ones who dashed the prospectors' hopes. It occurred frequently enough, the prospectors told Timothy, that "word got out, 'Prospectors, hey, the cloud is coming over. So, forget it this week." Junior Hicks told Timothy, "If you had a vacuum cleaner . . . you could have made money" simply by picking up the radioactive dust covering the ground after tests.[41]

The stories the Duchesne prospectors related to Timothy were common across the Great Basin. As tales spread about uranium prospectors striking it rich, general usage of Geiger counters increased, giving downwind citizens access to the necessary tools to gain a rudimentary sense of the radiation levels surrounding them. Agatha Mannering of Ivins, Utah, "became interested in the uranium boom of the day. We purchased a very expensive nucleometer and a geiger [sic] counter, and we went out prospecting constantly." She remembered "when this fallout came, we monitored that fallout every day,

Effect on Geiger Counters

Fig. 5. "Effect on Geiger Counters" illustration from the *Atomic Tests in Nevada* pamphlet published by the Atomic Energy Commission (March 1957) and distributed in the downwind region. From the collection of Preston Jay Truman.

sometimes two or three times a day, because we were anxious to get out and prospect. You cannot prospect with the high background radiation that was there at the time." She described to Carole Gallagher how "right on our doorstep, right on the threshold, on the nucleometer it would kick over on the second scale. And out in our garden area, it would almost take in the third scale. . . . This would last for days and days. . . . We were told not to worry and not to be afraid of it."[42]

Growing up near St. George, Diane Nielson remembered her father "had a Geiger counter that he would go and check things with. He would come in and be really upset because it was reading radioactive. Verbally upset. I remember my dad several times going outside and walking around with that thing and having it go ratta-tat-tat. We would tag along with him and watch it — it would go up. I do remember the dial going up and down and hearing it click-click-click." Her father told the children, "Clean off, don't get it on you, wash up good. Don't play in [the dust]." Years later, Diane Nielson recalled, "Of course, kids are going to go and do the opposite. We'd sneak out in it and dig in it and have a good time."[43]

While at the time they had no way of knowing what sort of danger the levels might imply, many other people also became concerned by the high readings, a concern the AEC's zone management program hastily sought to diffuse. When, in the spring of 1955, Cedar City residents "with Geiger counters . . . expressed . . . the belief that an extensive fallout had occurred," the AEC assured residents the amounts were "not enough to worry about." Several months earlier, the AEC had reminded uranium prospectors in southern Utah that "tests this year again will likely result in increased radiation readings" and prove "misleading in . . . prospecting activity." That increased radiation, the AEC assured, was "*not expected to be hazardous to humans or livestock or to have any effect on crops*" (emphasis added). In a 1957 version of its pamphlet "Atomic Tests in Nevada," the AEC printed a cartoon of a bowlegged, Geiger counter–toting cowboy, replete with checkered shirt, cowboy hat, canteen, and stubble. Staring wide eyed and with mouth agape at his clicking Geiger counter, the cowboy had a question mark over his head. Next to the graphic, the AEC used calm text to reassure readers who had experienced similar concern: "We can expect many reports that 'Geiger counters were going crazy here today.' Reports like this may worry people unnecessarily. Don't let them bother you."[44]

The cartoon did not set all minds at ease. Many citizens felt the government had been less than forthright in its explanation of recent livestock deaths, and the AEC's casual dismissal of high Geiger counter readings struck many as disingenuous. Official explanations for changes in the downwind region were proving inadequate for many citizens, who were insulted by what they perceived as the AEC's condescension, typified in the cartoon of the bowlegged cowboy gaping at his clicking Geiger counter. Cracks had appeared in the AEC's authority, and citizens increasingly turned to each other for information. The stories they shared—about strange headaches after a day in the fields or dust that withered vegetable plants—could now be shored up by the Geiger counter readings they either had heard rumors of or had witnessed themselves. Gradually downwind citizens were constructing their own narrative account of the effects of radiation exposure, an account that increasingly privileged local observations and knowledge over that of more educated and powerful outsiders.

The AEC was unconcerned with zone management in those areas where uranium was actively being extracted, even though numerous communities were in the path of severe contamination, often from both the uranium industry and the Nevada Test Site. These communities tended to be economically depressed and their inhabitants, people who were often indigenous or otherwise not white, lacked significant political power. "Managing" their perceptions of the uranium industry and nuclear weapons testing was irrelevant to AEC officials, because these people could not interfere with the AEC's activities.

The pattern of locating nuclear activities near poor, rural, and nonwhite communities was not limited to the United States. Russia established the Semipalatinsk site in rural Kazakhstan after test site planners, overlooking hundreds of thousands of residents and farms, falsely claimed the region was uninhabited. When Great Britain entered the arms race, its scientists opted to test their weapons in southern Australia's Maralinga region, which was populated by the indigenous Maralinga Tjarutja people. Uranium industries outside the United States likewise have tended to be located in arid regions, such as the Rössing Uranium Ltd. Mine in Namibia, that are inhabited primarily by indigenous peoples and farmers and ranchers. When sicknesses and deaths occurred in these communities, the government could easily overlook or dismiss them, because the affected people had little access to publicity or political or legal support. Many even lacked access to health care. This pattern has played out in numerous industries around the world, and in a 1978 lawsuit it gained a name—environmental discrimination. Time and again, researchers have concluded that communities populated by poor people of color are disproportionately likely to host toxic industry or waste.

Environmental discrimination is connected to a larger system called colonialism. *Colonialism* is the expansion of a nation's land, resource, and labor bases through the appropriation of the land, resources, and labor of the less powerful. The powerful do not want nuclear weapons tested in their backyards, so they have tested them in colonized, undervalued areas far from their nations' centers of power. Great Britain tested its bombs in what was then its colony of Australia. The United States needed uranium to build its bombs (and later to fuel its power plants) so it extracted the ore from the lands of sovereign Indian nations, nations that had been colonized by the U.S. government. Scholar Valerie Kuletz documents how "the uranium industry exploited the low visibility and lack of political power of the semi-sovereign Indian nations (reservations) to bypass [federal] environmental protection standards and job safety regulations . . . to ensure a high profit margin in the extraction, processing, and sale of uranium ore to the . . . scientific-military complex."[45]

Most of these rural and indigenous residents were familiar with the process of mining; yet in the more remote communities, particularly on reservations, they had very little knowledge of radiation or how it might result from the uranium industry. Residents were not blind to the intrusion of visible mine wastes into their surroundings. They noticed it particularly when it occurred via water, a precious resource in the arid region and one they could ill afford to ignore. As time passed, changes began to manifest in their crops and in the bodies of their livestock and the wild animals they hunted, allowing those people living downwind and downstream from the uranium industry to begin piecing together cause and effect.

In December 1995, Tommy James succinctly described the path of uranium contamination to Phil Harrison: "The waste was dumped over there. The water runs right through the waste, and we use the water. That is the same stream that was used for drilling in the mines. That is the same water the people drink, too. That is how it is with us today. And we plant our garden, but it does not grow big." James remembered how "years ago, before they did any mining in the mountain, when we planted the plants would get very big, such as watermelons and squash. It is the same in Shiprock; the farm produce are small [now]. The melons are small." Timothy Benally recounted a similar scenario: "The Navajo people say the [mine] operators went up into the mountains and pushed a lot of the dirt that contained some radiation or uranium off the side of the mountains and they were just scattered down below. When it rains and when it thaws in the springtime, a lot of the water washes into the riverbed and flows down into the stream and eventually comes out on the farms." Manny Pino described how at the open-pit Jackpile Mine in Laguna Pueblo, "there's two rivers, or streambeds, that flow directly through the mine. The Rio Paguate, and then . . . there was another tributary called the Rio Moquino, and they met in the center of the mine, and then they exit. . . . These two tributaries [flowed] downstream into another tributary called the Rio San Jose, which flowed into the Rio Puerco, which eventually flowed into the Rio Grande, the lifeblood of New Mexico."[46]

As essayist Joan Didion once observed, "Water is important to people who do not have it." Average rainfall in the Colorado Plateau lowlands is less than ten inches annually, and local residents make full use of that small amount. Visible contamination of the water supply by any process, be it erosion or the dumping or flowing of mine wastes, could not occur without attracting the local people's attention. As those who labored in the uranium industry began falling ill in the 1960s and, with alarming frequency, dying young in the early 1970s, some began to connect the sick miners to the poor crop yields in fields that had been irrigated with water contaminated by the mines. Activists such as Harry Tome and Manny Pino began the uphill struggle of educating their communities on radiation, a concept for which there was no word in their native languages. Memories of mine wastes entering the water supply and then the food supply became integrated with this new knowledge. "It's like a chain reaction," Timothy Benally explained years later. "The food you raise may have some radiation, and you eat it."[47]

Explaining that by-products of the uranium industry were radioactive did not always induce people to make the changes necessary to protect their health. As Manny Pino discovered, established practices could not be abandoned overnight, and in communities struggling to maintain their cultural identity and continuity of traditions, many were loath to alter their foodways. In 2005, Pino visited the Spokane reservation in eastern Washington, site of the open-pit Midnight Uranium Mine. He recalled, "The people were already being told, 'Don't eat the salmon from the Spokane River, because they're not fit for human consumption.' What do the people still do? Eat the salmon." He paralleled the experience of the Spokane to that of his own community. "It's like telling these people at Paguate, 'Don't grow corn, you know, because there's radioactive dust blowing.' Or 'don't dry your fruit in the open air, don't dry your meat out there.' This is a traditional lifeway, you know? These are traditional foods that people aren't going to give up that easy."

Awareness of the intrusion of radiation into the local foodshed came about unevenly in the downwind region. For most downwinders and uranium-affected people, catastrophic illnesses and deaths prompted questions about the causes, which led the questioners to reconstruct the contamination of the food chain and thereby identify the potential cause of illness. For some, such as Dave Timothy and the widows of many uranium workers, these illnesses occurred in the late 1960s and early 1970s and coincided with the emergence of new information about the potential health effects of radiation exposure, giving some of those who survived reason to ask questions in that early period. For others, such as Ivan Sidney, illnesses in their community and their family were simply isolated and inexplicable tragedies until new information or a new medical diagnosis put a pattern into focus for them, inevitably prompting questions about how radiation might have reached their bodies.

On a fundamental level, stories about the infiltration of radiation into local water and agriculture were simply variations on an old theme. For generations, the survival, comfort, and success of rural people in the American West have depended on their ability to function cooperatively, to share information as well as labor and culture. Knowledge about environmental factors, including the safety and abundance of local water, was integral to agricultural success and was thus shared with others in the community through story telling. Water folklore abounds in every region of the West, from such place names as Poison Creek and Clearwater to jokes about the control of sparse water supplies and folk prescriptions for provoking or anticipating rain. Wallace Stegner once noted "how often Shoshonean place names contain the syllable -pah: Tonopah, Ivanpah, Pahrump,

Paria. In the Shoshonean language, -*pah* means water, or water hole. The Pah-Utes [Paiutes] are the Water Utes, taking their name from their rarest and most precious resource."[48]

Such folklore does more than entertain or express the identity of a community; for generations, it has served as a spoken library of local knowledge, a narrative tradition that enables its users to live and raise food successfully in the region. In the words of essayist and farmer Wendell Berry, "Such a culture contains, and conveys to succeeding generations, the history of the use of the place and the knowledge of how the place may be lived in and used." This spoken history is a dynamic creation, as its relevance depends on constant retelling and updating when new information becomes available or environmental changes are observed.[49]

During the atmospheric-testing era and the first uranium boom, citizens in the downwind region had virtually no knowledge about the effects of radiation on plants and animals, and many were not even aware radiation existed. Expertise on the local environment and the process of raising food abounded, however. Rural residents knew that changes in their environment would affect their crops, so they paid attention to those changes. Years prior to learning the dangers of radiation, individuals living near uranium mines noted the intrusion of mine wastes into local water supplies. They filed their observations away in their library of local knowledge, and when their crops failed to thrive, they revisited the stories of water contamination by the mines. The Cedar City sheep ranchers took the same steps to deduce the cause of their losses. When all of their experiential ranching knowledge failed to explain the deaths and illnesses among their sheep, they concluded that the new contaminant they had observed in the region—dust from the test site—was to blame. Struggling to make sense of their cancer diagnoses, Dave Timothy, Ivan Sidney, and others relied upon their local knowledge to pinpoint the way contamination had reached them. No matter when individuals in the downwind region began to question the safety of the uranium industry or the test site, their participation in the local culture of food production aided their inquiry. Their membership in this culture invested their stories with emotional weight and rhetorical currency for other members of the culture who recognized the places, agricultural practices, and other cultural markers in the stories.

Even as this culture enabled its members to identify some of the changes being advanced by the atomic age, it was already beginning to erode. Agriculture alone was no longer enough to sustain many families economically, and young people were leaving their home communities in greater numbers to seek work and education elsewhere, creating a gap in the transmission of local knowledge to subsequent generations. Farmland and rangeland changed hands, and large commercial ranching and farming enterprises became the dominant producers of food and agricultural products. For these rural communities, as Wendell Berry observed, the fraying of local culture was both "a practical loss and an economic one." This loss, which was occurring regardless of nuclear activities, gained a new dimension when individuals considered it in the light of the nuclear contamination they learned about in later years. Not only was their way of life no longer economically sustainable, it had become poisoned—and by government negligence—rendering the loss of the local agricultural tradition and their loved ones that much more difficult to bear.[50]

{5} WRITING DOWN NAMES

As it turns out there were people in all those little communities—Mesquite, Bunkerville, Cedar, Parowan—that were doing similar things as what my mom was doing. They were writing down names, there were people tracking it.

Michelle Thomas, St. George, Utah[1]

Being downwind means you receive what the wind carries from other places, whether it is the scent of a neighborhood barbecue or the odor of a landfill. Folk wisdom reminds us that if an animal is downwind from a hunter, it is likely to smell danger—human scent—on the breeze. Downstream has a similar connotation. Wise campers refrain from muddying the water upstream from their campsite, thereby preventing contamination from entering their water supply.

The directional reference in "downwind" is also evocative. American society tends to connote power and value in spatial terms. Successful people are referred to as climbers who ascend metaphorical ladders to the upper classes and break through limits conceived of as ceilings. People whose quality of life is poor are said to be down-trodden or down and out, and when their circumstances worsen, they are said to be going downhill, presumably deeper into the lower class. The word "downwind" sounds even more ominous, when the thing upwind is a nuclear facility, particularly to those modern ears that have been conditioned by the accidents at Chernobyl and Three Mile Island to think of nuclear facilities as volatile and dangerous.

In the 1950s, living downwind of a government nuclear facility or a uranium site did not automatically cause alarm. Federal Civil Defense Administration (FCDA) materials regularly reminded citizens: "Radioactivity is nothing new . . . the whole world is radioactive." The risk came, authorities suggested, "if you are exposed to it long enough — IT WILL HURT YOU! IT MAY EVEN KILL YOU!" In the case of nuclear attack, the administration's 1955 "Facts about Fallout" pamphlet informed concerned citizens that "an ordinary frame house will offer some protection. It may cut radiation danger by about one-half. Get on the floor, away from doors and windows, or preferably go to a location with additional walls at the center of the house." For "almost complete protection," Americans were urged to construct bomb shelters for their families, preferably "[a]n underground shelter with three feet of earth above it." Then they were supposed to "stock it with food staples as Grandmother did. . . . Everything you need to live in it for a few days." The pamphlet followed the all-caps warning and bomb shelter specifics with reassurances: "Don't get discouraged, don't get panicky. . . . Americans are hard to scare. Of course, we are worried about the forces science has unlocked. We would not be intelligent human beings otherwise. But this problem can be solved, as others have been, by American ingenuity and careful preparation."[2]

Such FCDA messages were part of a campaign to prepare Americans for the possibility of a catastrophic nuclear attack while simultaneously reassuring them to go on about their daily lives. The problem, as historian Laura McEnaney describes it, was that "there was never any agreement among congressmen, defense planners, scientists, and civic groups about how a regionally, racially, and economically diverse population could be protected from atomic annihilation." Public bomb shelters were perceived as being too communistic, and they raised the awkward problem of segregation, which was still the law of much of the land in the 1950s. As nuclear devices grew steadily larger, the futility of attempting to shelter large numbers of citizens from weapons of unimaginable power became increasingly problematic. In order to marshal public support for continued weapons development, national leaders needed to convince the public that measures could be taken to save them in case of an attack.[3]

By the mid-1950s, the FCDA was placing responsibility for survival in the hands of the private citizens. All families were urged to construct their own private bomb shelter and to stock it with foodstuffs and survival equipment. This strategy saved the federal government the costs and cultural headache of constructing a network of public bomb shelters. McEnaney describes another benefit for national leaders: "The private home shelter could also perform the cultural work of the Cold War: it symbolized the superiority of a society of autonomous, property-owning individuals and strong families who had the capability and the choice to shelter themselves." Residents of large cities, who were disproportionately people of color, did not fit into this civil defense strategy; indeed, military experts were predicting that attacks would wipe out cities entirely and that society's only chance of survival was decentralization. That the first head of the FCDA, Millard Caldwell, was a well-known segregationist did not escape the notice of the National Association for the Advancement of Colored People, which attempted and failed to block his appointment. The white suburban family, defined "narrowly as husband-breadwinner, wife-homemaker, and several children," became the target of the FCDA's campaign. It was no accident: advertisers marketed consumer goods to these same citizens, and politicians courted their vote.[4]

In order to bring home their message, the FCDA televised a series of civil defense drills undertaken during nuclear tests in Nevada. At various distances from ground zero, it constructed single-family homes and stocked each with household items and mannequin families and fitted them with different styles of bomb shelters. Some had cars parked in their driveways; there was even a school bus filled with mannequin children. It filmed the homes during the nuclear weapons' detonations and examined the aftereffects. Intact mannequins and household goods in bomb shelters constructed according to specifications were highlighted as proof that the prepared family could weather an attack. "These tests were as much public relations morality plays about the dangers of family apathy as they were scientific experiments," McEnaney writes. "Viewed on television by millions of Americans, [these tests] defined who and what was endangered by the atomic age: families, homes, consumer commodities." The family of mannequins, and "the peril it endured, was both a warning and a reassurance: Americans had to fight apathy in order to spare the structural, material, and emotional enjoyments of postwar life from nuclear destruction." The white, two-parent suburban families depicted in FCDA propaganda made it clear who the government anticipated would survive to inherit and rebuild American society in case of an attack.[5]

Science-fiction writers reinforced this message. Widely read authors such as Ray Bradbury, Philip Wylie, and Stuart Cloete depicted white families rebuilding society after the destruction of nuclear war. Even when they criticized the political and social structures that led to the death of people of color, even as they depicted the ravages of nuclear weapons and criticized the nuclear arms buildup, these stories still helped white, middle-class Americans feel better about the new nuclear age. Survivors would be people like them, and the stories suggested these post-nuclear frontier families would form a more just and principled American society. Echoing Frederick Jackson Turner's half-century-old thesis that the best of American character came from the challenges of settling new territory, this fiction shored up the FCDA's propaganda. Patrick Sharp describes how the main character in Stuart Cloete's "The Blast" was "reinvigorated" by his "return to a savage lifestyle on the frontier." The character's "struggle with nature—including giant beasts mutated by radiation—helped him slough off his years of lethargy from his sedentary city life." Nuclear technology was depicted as both the poison and the cure: in an overwhelmingly technological age, filled with new threats and conveniences, nuclear apocalypse could erase the worst of the society that had created it (along with plenty of innocent bystanders) and clear space for the best of society (white Americans) to create a more moral nation.[6]

Nuclear war could even bring the family closer together, FCDA planners suggested. While mothers stocked the emergency pantry and fathers used some old-fashioned muscle to dig the family shelter, they could look forward to time playing cards with the kids if the bomb ever dropped. The FCDA materials implied that the prepared family had nothing to fear and even could look forward to a little time off work. This silver lining was underscored by the results of a Princeton University psychology experiment, which subjected a family of five to two weeks of shelter life. With the help of a tranquilizer administered to one of the children, and having whisky available for the parents, the family weathered the time without significant psychological damage and was said to have emerged with a "very positive attitude." The researcher noted the shelter stay had given the father an "opportunity to get to know his children better," giving him a "new-found respect for the entire family."[7]

Along with Americans in the rest of the country, Southern Utah and Nevada residents diligently built bomb shelters in their basements, and they and residents of nearby reservations participated in civil defense drills. Because the bombs detonated at the Nevada Test Site were "tests," supposedly controlled by scientists and military experts, and nearby uranium mines were "workplaces," the radioactivity associated with these aspects of the Cold War was made to seem a non-issue. Those people living near the NTS never bothered to use their bomb shelters after nearby detonations.

The contradiction did not escape St. George, Utah, resident Irma Thomas. Her daughter Michelle remembered coming home from school to find her mother in the basement, "boarding up the tiny window in our little food storage room. She was doing what the government had instructed Americans to do: building a makeshift bomb shelter in the event Russia launched an atomic war." Decades later, Michelle wrote about watching her mother "stacking cans of food on a shelf in the bomb shelter. 'This makes no sense at all,' she growled. 'I'm readying a bomb shelter to keep my family protected from [Nikita] Khrushchev's bombs and our own government is dropping bombs nearly every day on our heads and telling us we're safe!'"8

Despite her apprehensions about nearby test site activities, Irma Thomas kept her concerns quiet initially. Citizens in the downwind region were continually reassured they had nothing to fear from testing activity, while politicians and newspaper headlines daily stoked fears of a Russian nuclear attack. "I spent much of my early childhood terrorized by the name Khrushchev and his bombs," Michelle Thomas later reflected. "What if I wasn't home [to use the shelter] when he bombed us? Where could I be safe?" Claudia Peterson remembered, "My biggest fear was that we were going to have to live in the potato pit when they bombed. . . . It wasn't a bomb shelter, it was a potato pit. And I thought, 'Oh, there's spiders in there and we're going to have to hide in there and live there for two weeks.'" Like many children of the baby boomer generation, Claudia remembered atomic bombs being her "biggest fear as a child. But it was more of what the Russians were going to do to us, not what [our] people were doing to us."

Local representatives, press releases, and public relations material from the Atomic Energy Commission and Public Health Service, meanwhile, urged downwind residents to feel pride in the tests mushrooming on the horizon and even to be spectators. Sheldon Johnson of

Fig. 6. "Procedures for Flash" illustration from the *Atomic Tests in Nevada* pamphlet published by the Atomic Energy Commission (March 1957) and distributed in the downwind region. From the collection of Preston Jay Truman.

St. George felt he and his neighbors "were important people, because we were right in the middle of science's most advanced knowledge." He described finding the nuclear tests "exciting because we could show the principles of sound waves being slower than light. We were up at the airport and watching the flare, and you could see the sky light up, right down through there . . . and then we would count how long it would take before the sound wave got here."

Quentin Nisson, the mayor of Washington City, was invited to the test site to watch a detonation with other local mayors in the early 1950s. "I got the privilege of getting a letter, and if I rode down to Vegas, got on their bus, a bus tour was free from there out to Mercury [the town constructed on the premises of the test site to house workers], but I had to go down about five times." Nisson had to return five times because every time he made the trip, the planned nuclear test was canceled due to wind direction. "You know they wouldn't shoot it if the wind was blowing towards Vegas," he explained in 2005. "It had to come this way." By "this way," Nisson was referring to his hometown of Washington City, located east of the test site and near St. George in southern Utah. Eventually, the wind turned east, and Nisson was able to witness and photograph a test. "We had to wear those doggone dark glasses, so thick, you know, and that was the reason that I didn't get the initial flash. They told us, you know, to brace yourselves. But I found out one thing, though—it isn't like a stick of dynamite, it's a pressure, when it goes. The air actually gets to you almost before the sound. It's a funny experience. I did get oriented enough to get [a picture of] the mushroom going up."

The public was encouraged to view the "history-making" shots from their communities and to wear sunglasses to protect their eyes. Loa Johnson remembered how "teachers took their students out to see history made. It was going to be an explosion at such and such a time, so they'd take them outside to watch, to see what they could see." After Loa lost her daughter Sybil to leukemia several years later, she remembered these school excursions with painful irony. "They made a big deal out of telling us that it would harm no one or nothing. Of course, it couldn't do any harm at all. Of course, we didn't stop to think that they waited until everything was just right, that the breeze was blowing in this direction rather than toward Los Angeles or Las Vegas. And I can see their reasoning, but it wasn't very fair."[9]

By publicizing the tests to cultivate the interest and pride of the downwind residents, the AEC unwittingly enabled residents to develop a rudimentary understanding of test site protocol: nuclear tests only went forward when the wind blew east. For those locals who found this prerequisite troubling, a stern message lurked beneath the celebratory publicity. By "[s]ubtly questioning the patriotism of those who might criticize the Nevada tests," Frederick Schoemehl suggests, "the AEC made it clear national security would be jeopardized if residents did not demonstrate 'complete' acceptance of the program." The AEC reminded downwind residents, "Your best action is not to be worried about fallout. If you are in a fallout area, you will be advised." Most downwind residents took this reassurance to heart initially, yet as the years passed, a mounting toll of inexplicable losses made it difficult "not to be worried."[10]

Clusters of miscarriages were some of the earliest patterns of illness to manifest downwind. Stories about these family tragedies most likely circulated quietly at the time, but there is no question they were shared, with verifiable details becoming secondary to the emotional impact of the pattern. Given a few decades of retelling, it is not surprising that by 1978 a *Deseret (Salt Lake City) News* article confidently offered as fact the rumor that "all the girls who were school-age during the atomic testing period later suffered miscarriages during their first pregnancy." Ina Iverson, who lived in the town of Gunlock, Utah, as a young woman, recalled that "all of a sudden everybody we knew or somebody they knew or were related to was having miscarriages at five or six months along, or their babies dying on them, and I said something has got to be going crazy here. . . . I thought they put something in the water!"[11]

Ena Cooper, who lived in the small community of Indian Springs at the edge of the NTS, recalled one test cloud "rolling by our area. There were three women pregnant in the trailer court at that time and all of their babies were born dead. They just deteriorated inside their mothers." Martha Bordoli Laird, who lived on a ranch in Twin Springs, Nevada, north of the test site, gave birth to an infant whose body "from the hips down the legs [was] all shriveled up and black. It lived for a couple of hours. . . . That was a seven-and-a-half-month baby. It was 1953." Laird's inclusion of the date 1953 is hardly incidental. Stories of miscarriages and birth defects abounded after the well-remembered tests of the 1953 Upshot-Knothole series; in fact, so many were told that the year became a focal point in the memories of many downwind residents, even those who were not old enough to have known about it at the time. Some narrators even attribute particular health problems specifically to Shot Harry, the most notorious test of the 1953 series.[12]

Waves of birth defects appeared along with the miscarriages, rippling through communities too small to ignore them. Glenna Orton of Parowan, Utah, remembered "seven retarded children born through that time [in the mid to late 1950s], more than we had ever known in the whole history of Parowan." Layne Johnson was one of nearly a dozen children born with Down syndrome in St. George the year following the Upshot-Knothole series. "You have to realize that the town at that time was a very small town . . . only thirty-two hundred people, thirty-five hundred. . . . Now we're up to seventy thousand or seventy-five thousand people," Layne's father, Sheldon, told me in 2005. "So there were very few people with Down syndrome even in the whole town, except this group of them that were born at that exact time. I know that isn't science, but it was pretty obvious there was something going on. The interesting thing, we got them together and started this classroom unit [for Down syndrome children], and finally got the unit going. And then, three or four years later they thought, well, there will be other ages coming along, but there weren't any."[13]

On the heels of these initial patterns of death and disability, leukemia began to appear in the downwind region with increasing frequency, particularly among children. "We had seven young people die of leukemia in two years around 1957, and we had never heard of a leukemia death before that time. That's when I think people began to be afraid that it was the radiation that was causing the problems," Glenna Orton later told Carole Gallagher. Orton likely knew the names and families of each child who had succumbed to leukemia and of each child who was born with some form of mental handicap, but it is interesting that she remembered exactly seven of each group. Folklorists have noted that numerical repetition adds rhetorical potency to a story, and some numbers resonate more for us than others do. The number seven has special significance in human narrative. From fairy tales to spiritual texts from across the world, it appears constantly: seven days of creation, seven brothers for seven brides, seven deadly sins, and seven wonders of the world. In Glenna's story of life downwind, seven mentally handicapped children and seven childhood leukemia deaths have a similar potency of meaning.

When Martin (known as Butch), Martha Bordoli's son, was diagnosed with stem cell leukemia in 1955, many of his doctors had never treated the disease. Butch died ten months later of "severe gastrointestinal hemorrhage" brought on by the leukemia. By then, Martha and her neighbors were well aware of the nuclear tests to the south and were thoroughly unnerved. Martha described the tests for Gallagher as beginning with a "big ball of fire in the sky . . . the flash was so bright it was like the sun would come out right in your face, and then it wasn't long after that you could hear it coming." The sound that followed the bright ball of fire "was louder than an earthquake and you could feel the ground rumbling. Then when it got to you, you had the shock." This shock wave was powerful enough to shatter windows at the Bordoli ranch on one occasion. "We lived right under it," Martha told Gallagher. "I have pictures . . . that prove it. There's a big cloud coming in."

The Bordoli ranch was beneath the debris trajectory that meteorologists Robert List and Lester Machta identified in their 1952 report to the AEC. They noted how "time and again, often in direct contradiction of the weather patterns, low-level debris [generated by nuclear tests] would track north, following the ridges to the northern Nevada town of Elko." Martha recalled one cloud of "smoke so black, just like it comes out of a train." The cloud drifted north after the aptly named 44-kiloton nuclear test Smoky was detonated atop a 700-foot tower on 31 August 1957. That same day Martha Bordoli made a trip into the nearby town of Tonopah. She stopped by the *Tonopah (NV) Times-Bonanza* newspaper office, owned and run by

Robert Crandall and Minette Crandall. Martha remembered Minette holding a Geiger counter to her hair. "Well the needle went right over to the pin, click, click, clicking. . . . There was one of those smart-ass [AEC] monitors over there, and he said . . . "Just wash your hair and it will be all right."[14]

Martha Bordoli and the Crandalls put little stock in the AEC monitor's advice. Subsequent to other tests, Bordoli had observed strange eye problems and white spots on the hides of her cattle and rashes on her children's skin. The Crandalls had always been ill at ease with the testing and routinely questioned the government's assurances of safety in their small-town newspaper. When Butch Bordoli died of leukemia in 1956, Robert Crandall wrote about the family in the *Tonopah (NV) Times-Bonanza*, suggesting they lived in an area of high fallout. In an effort to convince his readers of the dangers, Crandall contacted Nobel Prize–winning chemist Linus Pauling, who guessed the lifespans of Tonopah area residents might have been shortened by three months due to radiation exposure. Crandall multiplied the guess by the number of area residents and ran the story under the headline "Local Citizens 'Give Up' 1,000 Years." The following May, as the Operation Plumbbob test series got under way to the south, Crandall framed an exchange between area residents and the AEC in the pages of the paper.

> Aren't we folks who live around "uninhabited areas" and who like to go hunting, fishing, camping and prospecting in out-of-the-way spots, betting the short end of the odds when you refuse to fire an atomic shot when the wind blows away from us? And then, inevitably, as it must when the AEC finds itself on crumbling ground, it turns questioner to demand: "Do you want us to halt the Nevada tests, and let our enemies win the atomic race for survival?" No, we don't say that at all. But what we do say is this: "Alright AEC, go ahead with your tests. But if you must shoot craps with destiny, first throw away the loaded dice."[15]

The newspaper's early and persistent questioning of the AEC's claims offered local residents a counternarrative to the patronizing rhetoric of the government. As more women miscarried and the stories of strange new diseases eddied from sickrooms to kitchens to stockyards, Martha Bordoli penned and circulated a petition among her neighbors in 1957: "We, the residents of the area immediately adjacent to the Nevada test site, which the AEC has designated as 'virtually uninhabited' feel that the health and welfare of our children and ourselves have been seriously endangered." Playing into the popular perception of area residents, the petition affirmed, "We are not excitable or imaginative people, coming from rugged ranch families, but neither are we without deep feeling for each other and our children. Having with our own eyes seen tragedy strike among us from 'radiation associated' causes, we feel that official recognition of our plight is long overdue."[16]

The language Martha Bordoli used to address the federal government in her petition letter came from the same vernacular that the sheep ranchers used when testifying opposite the AEC in federal court in Salt Lake City. While admitting a lack of scientific expertise, Bordoli and the sheep men invoked both rural practicality and humanity and spoke of radiation effects observed "with our own eyes." In their humble way, while protesting the sacrifice of their homeland and resisting the dismissal of their losses as insignificant, they stood their ground against a bureaucratic entity bent on global supremacy. Casting themselves as hard-working, family-oriented people, these downwind residents reminded federal authorities that they were not troublemakers or agitators. They were patriotic Americans from the heart of the nation, and they neither asked for grand settlements nor wanted to disrupt the work of national security. They sought instead remedies the government could have easily provided: recognition of their losses, respect for their ability to see those losses as connected to nuclear activities, and the decency of warnings they could use to protect their families and livelihoods.

Lewis Strauss, then chair of the Atomic Energy Commission, responded to Martha Bordoli. He wrote and suggested she take the advice of President Truman: "Let us keep our sense of proportion in regard to the matter of radioactive fall-out . . . the dangers that might occur from the fall-out in our tests involve a small sacrifice when compared to the infinitely greater evil of the use of nuclear bombs in war." Strauss continued, "The risks from the current rate of nuclear testing are small, exceedingly small in fact when compared to other risks that we routinely and willingly accept every day. I believe the conclusion one must inevitably reach after balancing all factors is this—we have the choice of running a very small risk from testing or a risk of catastrophe from a surrender of our leadership in nuclear armament which has been, we believe, the deterrent to aggression since 1945." Concluding his letter, Strauss added, "I have a wife, a son, and three happy young grandchildren," and he emphasized that he could not support testing "if I thought the welfare of my family threatened more by fallout than by possible nuclear war."[17]

Strauss's suggestion that the Bordolis keep their "sense of proportion" must have come as a slap in the face, particularly alongside the intimation that their son's painful death was a "small sacrifice." His reassurances rang hollow for Martha Bordoli, who likely guessed the chairman's wife, son, and "three happy young grandchildren" lived nowhere near the test site. The Bordoli family put their ranch on the market and moved to Carson City. "We knew the government wasn't going to do anything, and my husband wasn't well," Martha told Carole Gallagher in 1988. "We didn't want to expose the two girls to any more radiation. We sold our ranch for a song." Shortly after their departure, Martha's husband lost his battle with cancer. He was "only fifty-something when he died."

Around the time that Butch Bordoli died, Cedar City toddler Sherry Millet began to suffer from nosebleeds and strange bruises. Her mother, Kay, recalled: "Of course, the doctor didn't know what was causing it. . . . They weren't that familiar with leukemia. It was something he said was really rare. He had only known of one case of leukemia during all his doctoring years here in southern Utah." After examining Sherry, the family doctor instructed Kay to take her daughter "to Salt Lake first thing in the morning." The hospital staff in Salt Lake City told Kay to leave Sherry overnight. When she returned the next morning, she found her daughter "all pink. They were experimenting with all these new drugs at this time."[18]

Decades after Sherry's death, Kay reflected, "That's what made me think we were just guinea pigs, they knew this was going to crop up and they were coming up with drugs they were trying to test on patients that would come in." Millet and her family took Sherry on the long drive to the hospital in Salt Lake "every week for treatment and transfusions, whatever they thought she needed." In the hospital waiting rooms, the Millet family discovered they were not alone. "There was a hundred little kids with a rare disease called leukemia going to that hospital for treatment," Kay recounted to Carole Gallagher. "We didn't know until we went up there that there were that many kids suffering with it. All of a sudden this rare disease wasn't rare anymore. It was just raging inside of Utah." She met families who had come all the way from the neighboring states of Idaho and Wyoming to seek treatment for their children. "It wasn't only leukemia, there was Hodgkin's disease and brain tumors all of a sudden. One day I just sat down and thought, 'All these people, there must be something causing this for everybody to be getting cancer.'"

Sherry succumbed to leukemia two years later at the age of five. As the parent of a terminally ill child with a rare and poorly understood disease who required constant treatment by specialists in a hospital located five hours away, Kay Millet spent ample time in waiting rooms with other frightened parents and children. Many were battling the same rare and poorly understood diseases with the same experimental treatments, all grasping for answers and finding their doctors could offer them little information or solace. The kind of knowledge that emerges from such an experience is impossible to quantify or to translate into epidemiological data. For a parent like Kay Millet, no scientific report was necessary to tell her what she had seen with her own eyes: "all of a sudden . . . [leukemia] was just raging" in her region.

Sheldon Nisson of Washington City lost his battle with leukemia in July 1959, shortly after his thirteenth birthday. His mother, Helen, could not "remember anyone [before Sheldon] that I knew of that died of leukemia, yet our little school had seventy-five kids, and Jesselyn Turner died just a few months before Sheldon, and they both had leukemia, and there's a little baby out to Enterprise about the same time died, and then Carleen Hafen in St. George died about the same time." One leukemia death in a school of seventy-five children would have been a tragedy, two a strange coincidence. For Helen Nisson and other residents, four young children in a rural region succumbing to rare diseases around the same time made for more than a tragedy or a coincidence; it made a pattern.[19]

The deaths reverberated through the children's peer groups. Sheldon's uncle Quentin described how Bob, his son, and Sheldon "played together, they played down here [gesturing to his backyard] in the old turkey shed. They took a [model] plane in there, and they built it so big they couldn't get it out." Quentin reflected that Sheldon's death had "been hard on Bob, and I think it's bothered him, because 'if Sheldon got it, I wonder if I'm gonna get it.' You can only naturally think that, you know."

In 1964, doctors diagnosed twelve-year-old Sybil Johnson of Cedar City with leukemia; she passed away in May of the following year. Her mother, Loa, recalled Sybil "had a lot of friends. There were, I guess, a hundred of the young kids from her school, you know—they weren't all close friends—that were conscious of her [illness]. She had this little group of girlfriends that held a prayer meeting every morning before school." For these children of the atmospheric-testing era, witnessing the premature deaths of their peers profoundly affected their sense of the future. Mary Dickson reflected, "As a child growing up [in Salt Lake], there were kids on my block who died of cancer and leukemia. We just thought that was what happened to everybody. That people got cancer, they died."

In the late 1960s, relatively young men, most in their late thirties and early forties, began to fall ill in the small Navajo community of Red Valley, near Shiprock, New Mexico. The men complained of severe pains in their sides or chests, breathing difficulty, and sometimes skin problems. When they sought medical care at the Shiprock Indian Health Service hospital, Peter Yazzie, Todacheenie Benally, Billy Johnson, and others received diagnoses of advanced-stage cancers, most frequently of the lung. In the early 1970s, these ailing men succumbed, one by one, often leaving behind wives and numerous children. Their deaths reverberated through Red Valley, and local tribal leader Harry Tome realized a pattern was beginning to emerge. "Nine times out of ten they worked in the uranium mines," Tome related years later. "I began to make some inquiries and talked to doctors. They said, 'yes,' there are studies that show it is harmful." The harm, doctors told Tome, had been inflicted via radiation, a word for which no equivalent existed in the Navajo language. It was explained to Tome as "dangerous smoke you breathe in."[20]

Phil Harrison's father, one of those Red Valley men, fell ill around 1970. The elder Harrison had begun working in the uranium industry twenty years earlier and had raised a family in the vicinity of its mines and mills. "He was patriotic," Phil recalled in 2008. "I think he was workin' there to take care of us, to have food on the table, clothes. That was his intent. So was everybody else [that worked in the industry]. But the government never told them about what was gonna happen. [The workers] never knew. I worked almost five months in a mine with my father." Fresh out of high school, Phil had enlisted in the military and was completing his basic training in Denver, Colorado, and preparing to go to Vietnam when he received word his father was succumbing to cancer. Phil's commanders placed him on reserve duty, and he returned home to Cove, his family's village in the Red Valley area. His father had "just turned forty-six. His birthday was December 27, and he died on January 11. Forty-six years old." Home again in Cove, working to support his mother and siblings, Phil began to hear stories of the other cancer victims. "All of a sudden, sixteen miners at an early age died. That's when our community, Red Valley, came together to say, 'You know what's going on, ladies and gentlemen? Our fathers are dying.'"

Fifteen years prior, cancer was so rarely diagnosed among the Navajo that medical researchers surmised the group might possess such a condition as "cancer immunity." By the early 1970s, it was amply clear to the residents of Red Valley that no such immunity existed. The cancer-stricken uranium workers provided an easily observed pattern for their communities, sharing as they did a common work history and a common diagnosis. Families who might not have connected their loved ones' illnesses to a larger pattern often did so once they encountered other families in waiting rooms at distant hospitals, just as Kay Millet and other test site downwinders had. Red Valley resident Fannie Yazzie told journalist Judy Pasternak how she raised money by selling sheep and pawning jewelry to finance her husband's move to the Albuquerque Veterans Administration (VA) hospital. "There, she met nine other Navajo women with nine other husbands who were dying of cancer." Every man had worked in the mines. Similar patterns emerged in uranium communities elsewhere in the region. Locals in Marysvale, Utah, began referring to the cemetery as "uranium hill" and "cancer hill."[21]

Around the same time, another group of men, roughly the same age as Phil Harrison's father and his counterparts, began to receive cancer diagnoses. These individuals had also worked on the front lines of the Cold War, and like the uranium workers, they had not received any indication from their employers that their health was ever in danger. Robert Carter was a teenage soldier in 1957 when he and his U.S. Air Force squad were stationed at the Nevada Test Site to witness the 74-kiloton nuclear test Shot Hood. Part of an effort various military branches conducted during the 1957 Operation Plumbbob, these soldiers were present at the test series so they could be conditioned to the psychological and tactical elements of nuclear warfare. By the early 1970s, cancer and leukemia were sweeping through their ranks.

"When the countdown came close I was scared to death," Carter told Carole Gallagher in 1988. The blast knocked Carter and his fellow soldiers "forty feet into the mountainside . . . I felt elbows, I felt knees, I felt heads banging . . . I had a bloody nose. . . . After the shot my coveralls were cracked and burned, there was so much heat." After the initial blasts of the Plumbbob tests, soldiers were ordered to march toward ground zero through the rubble of the buildings that had been constructed to test civil defense designs. They passed melted mannequins in military uniforms and cages of test animals that had been confined for scientific experiments and the testing of protective clothing. Robert Carter and other soldiers, including Israel Torres and Jim O'Connor, subsequently related, with great anguish, memories of seeing humans confined to cages or bunkers near the site of blasts, but no corroborating documentation has been found. As ghastly as the idea is, it is not out of the realm of possibility. The Department of Energy publicly admitted to more than four hundred human radiation experiments in its 1995 report *Human Radiation Experiments Associated with the U.S. Department of Energy and Its Predecessors*, which was released as part of Energy Secretary Hazel O'Leary's efforts to make the DOE's history and actions more transparent in the early 1990s.[22]

Many of these soldiers' memories cannot be verified, and it is reasonable to assume that some were exaggerated or imagined or possibly by-products of post-traumatic stress disorder, which has been studied and diagnosed among atomic vets. Post-traumatic stress disorder has undoubtedly influenced the narratives of civilian downwinders as well. Having survived a trauma or traumas too painful to reconcile, individuals often attempt to express that pain in stories, which may or may not conform to others' notions of historical truth. While it is important to take all stories with a grain of salt, those that are told repeatedly need to be examined closely. Even if they are not historically accurate, they have impacted the perceptions of all who told and heard them, and research often bears out the truth of seemingly unbelievable stories.[23]

Ted Przygucki was a master sergeant in the army and stationed at the Nevada Test Site in the early 1950s. He recalled, "The officers in charge, they said that all the safety precautions were taken to prevent anything from happening to us." Przygucki was diagnosed with cancer of the larynx in 1976. He and fellow veterans of the test site began to refer to themselves as "atomic vets." Przygucki's peers told him "that when they go for a check-up at the VA, and they tell [staff] they are atomic veterans, they just give them a normal physical. Then they call in a psychologist to try to talk them out of the idea that they are sick. . . . When I told them I was exposed to radiation, it was 'so what.'"[24]

It is difficult enough to receive a cancer diagnosis for oneself or a family member, difficult enough to survive the ravages of cancer treatment, difficult enough to learn to live without a loved one, breadwinner, parent, or child. To learn that case of cancer may not have been a random tragedy, that you or your loved ones were exposed to toxic pollution by a government that claimed to protect you but failed to warn you of the danger, is another level of difficulty entirely. Many citizens in the region downwind of the test site and uranium industry maintain the cancers of their loved ones were the will of God or simply bad luck. For others the pattern of loss became impossible to overlook. Some stored up the lists of cancer victims in their heads or their prayers. Others started writing down names.

Michelle Thomas remembers that in the early years of testing, her mother "just mumbled" about her mistrust of the test site and did what she could to limit her children's exposure. Irma Thomas "knew that it was dangerous, but she was willing to give them the benefit of the doubt, or maybe not." While Irma had no way of knowing the extent of radiological contamination occurring locally, she felt certain it was present, and she worked to protect her family's health the best way she knew how. In 2013 Michelle remembered:

When I was little I'd go out with her to take down the laundry. Everybody had a clothesline in those days. We didn't have a dryer. She'd let me hold the basket and collect the clothespins and I felt so special. And there were some days when she'd say, "You can't go out today, Shelly, its 'hot.' It's dangerous." And I was heartbroken. Then she'd get this big wide-brimmed hat and this canvas zip-up coverall outfit and rubber boots and tie a dish towel over her face. It was like a homemade 50's hazmat suit. I'd sit at the kitchen window with my face pressed up against the window, watching, so sad I couldn't go out, and she'd take down all the laundry and shake it out. She'd take off her suit at the back door and rewash the clothes twice in hot, hot water, and then she'd hang them to dry on the line in the basement. Our neighbor Mary Webber said, "Irma, you are nuts." And Mom says, "Well, I hope to be alive nuts. I don't want my kids sleeping on those sheets."[25]

In 1968, Irma Thomas's close friend Helen Reichmann, who lived across the street, died of cancer. "Everybody can remember the person or the situation [that led them to say,] 'I can't any longer believe that this is safe,'" Michelle reflected later. "I think Mom was functioning up until Helen." While Irma harbored strong suspicions about the risks of testing, "because so many little babies, children had died, leukemia and stuff," her daughter believes Helen's death "was the first time that I think it got so overwhelmingly personal that she could no longer not be openly outraged and start writing letters and going public." Many downwind and uranium-affected people describe a similar moment, a particular loss that changed a person's opinion about the risks of radiological exposure. "It's interesting to see people become members of the club," Michelle observed in

2005. "To go from 'There's no correlation' to 'Oh shit, I just lost my dad and my brother.'"

The shift Michelle describes, when "people become members of the club," represents a critical transformation by downwind residents and uranium-affected people. Most people are raised and conditioned to believe authority figures—government representatives, members of the military, scientists, doctors—particularly in times of crisis, such as wars, when citizens crave reassurance and patriotism takes on an element of blind trust. No matter how fervent a person's patriotism or respect for authority, however, this sort of trust is rarely absolute. If enough evidence arises to undermine the authority of these figures, citizens will revoke their trust. The moment at which this revocation occurs is different for everyone. For Irma Thomas, it took place in 1968, when, in her daughter's words, "it got so overwhelmingly personal that she could no longer not be openly outraged. . . . That's when she started the chart."[26]

Without access to nuclear science or formal epidemiology studies, Thomas developed her own method of tracking downwind illness. On a large piece of paper, she sketched out a map of her neighborhood. For much of Michelle's childhood, the map hung on the dining room wall. It had "square boxes drawn on it, rows and rows of square boxes. Each box represented the home of our neighbors within a three-block radius. Each time someone in the home was diagnosed with a disease—cancer, Down's syndrome birth, et cetera—Mom would draw an X on their house." When Michelle was in high school, doctors discovered precancerous cells on her ovaries, and she underwent surgery to have one ovary removed. Several years later, Michelle, an aspiring dancer, was diagnosed with a debilitating autoimmune disease that would rob her of much of the function of her legs. "Mom drew an X on our house."

A transcription of oral, local knowledge about disease patterns, Irma's map was a powerful visual tool when she discussed fallout-related illness with her neighbors. While many were uncomfortable with the housewife's increasingly vocal criticisms of the government and the military, they were attending the same funerals she was, and the homes of their family members were beginning to appear on her map. When Irma began inviting the media to her home to talk about downwind illness, the map proved equally effective in impressing journalists, and she invariably brought them to stand on the front porch. From the stoop of 375 Tabernacle Street, Thomas directed every reporter's attention to each home on the block that had been visited by health problems.

Many scientists and government officials dismissed Irma's map and lists as nothing more than the paranoia of a housewife without formal scientific training. Not so, claims medical sociologist Phil Brown; to him, maps such as Irma's can offer substantial evidence of very real patterns. He coined the term "popular epidemiology" to describe the work of creating those maps after studying the organizing efforts of residents—many of them housewives and mothers—of Love Canal, New York, in the 1970s. These residents observed and mapped a pattern of disease in their communities, a pattern eventually connected to massive quantities of toxic waste discarded by Hooker Chemicals Company.

Popular epidemiology, or the science Irma Thomas practiced, "is the process by which laypersons gather scientific data and other information and direct and marshal the knowledge and resources of experts to understand the epidemiology of disease," in Phil Brown and Edwin Mikkelson's words. This citizen science is more than simply a lay version of traditional epidemiology, the scientists argue. Popular epidemiology "goes further in emphasizing social structural factors as part of the causative chain of disease . . . involving social movements . . . utilizing political and judicial remedies, and . . . challenging basic assumptions of traditional epidemiology, risk assessment, and public health regulation." Where professional epidemiologists may write reports or make recommendations to government agencies, popular epidemiologists will call their neighbors,

Fig. 7. Home of the Irma Thomas family, 375 Tabernacle Street, St. George, Utah. Photo by the author, 2005.

speak up at local meetings, write letters to elected officials, and file lawsuits.[27]

As Irma Thomas and other individuals began observing clusters of illness in downwind communities, connecting those illnesses to lifestyle factors and physical geographies, and recording that knowledge in narratives, lists, and maps, they became practitioners of popular epidemiology. While many eventually used this evidence to get the attention of journalists, scientists, and politicians, first and foremost, they used these charts and lists to alert their families, friends, and neighbors. The list makers were becoming storytellers. Drawing on community relationships, the visual aid of lists and charts, and the potent rhetorical tool of local knowledge, these storytellers often persuaded others to begin keeping their own lists. Myron Glazer and Penina Glazer suggest these early activists "were on the frontier in learning that Cold War policies had created serious damage in their immediate environment, and they undertook distinct forms of organizing in their communities to demand recognition and remediation for long submerged and unarticulated problems." The Glazers argue these activists found some success because they "knew the local culture well and could employ a vocabulary that spoke as neighbor to neighbor. They could not easily be dismissed as outside agitators or troublesome radicals by those defending the status quo."[28]

Loa Johnson remembered how after her daughter Sybil's death from leukemia, "Irma Thomas . . . came to Cedar [City] . . . to talk to us about [testing]. She was quite upset. But we weren't upset. Because we hadn't paid much attention to it at that time." Thomas convinced Loa and her husband, Blaine, that Sybil's death was no random tragedy but part of a pattern of health problems that could be traced back to nuclear testing. By the time Irma left, Loa had agreed to become a gatherer of names. "My phone rang constantly for several days, people who had lost or who had been very seriously damaged, like losing all their income and so on," Johnson recalled. "I had great long lists of people. We lost twenty-one people within a two-block radius of our house from leukemia in that little neighborhood." Her list making led Loa Johnson to the conclusion that fallout "seemed to land in pockets. There were about four different areas in Cedar that were very highly hit, hard hit by [illness]. It just hit neighborhoods. I guess it's in clouds, you know, and landed in certain areas."

Residents in the downwind and uranium-producing regions began to describe patterns and clusters of disease observed in their personal geographies—unique, interconnected networks of families, coworkers, friends, and acquaintances and recreational activities and neighborhoods—and they began to observe the ways these patterns and clusters could be connected to nuclear testing or uranium extraction upwind. In doing so, they created narratives about experiences they had witnessed on personal, local, and regional levels. In describing their experience for themselves and others, they discovered its collectivity, and a shared identity and a shared narrative began to evolve. Although the terms "downwinder" and "uranium-affected person" did not come into common public usage until the late 1970s, the identity they described began to develop much earlier, and it has continued to emerge decades later.

Narrators began to describe patterns of disease linked by various factors of interdependence. A cluster of disease in a family might be described as crossing generations, then connected in the next sentence to a cluster of diseases in a school class, and then linked to a group of illnesses or cancers in a neighborhood or among individuals who labored at the same kind of work, like uranium mining or ranching. Narrators also referred to clusters of specific diseases, such as the leukemia cluster Kay Millet discovered in the hospital waiting room in Salt Lake and the lung cancer cluster Fannie Yazzie found in the VA hospital in Albuquerque.

Occupying as they do an imagined space, one dictated by physical proximity as well as the geographies of emotional attachment and genetic ties, these clusters and the stories describing them grow more extensive with the passage of time. The larger the pattern, the harder it is to dismiss as coincidence or bad luck. Because the spaces they occupy are largely universal—every human has emotional frameworks connected to notions of family, a peer group, and a neighborhood—the stories of these clusters of illnesses and deaths have the potential to resonate with any audience willing to listen.

Disease patterns, or clusters, are usually recounted as narrative lists, offering the name and relationship of the affected individuals and some description of their diagnosis or illness. Rhetorically recounting a list is an effective tool for convincing the audience of the breadth and seriousness of the health crisis being described. Persuading the audience of the severity of the tragedy at hand is almost always secondary to the storyteller's primary purpose in recounting the names of those lost—that is, bearing witness and keeping alive the memory of those who have died.

By far the most common pattern recounted in stories of life downwind is also the most intimate and centers on the clusters of disease that have occurred in the narrators' own families. No matter how critical the audience may be of the connection between the described diseases and nuclear contamination, it is difficult to listen to people say they have buried their child, their parents, their siblings, and their grandparents, then have them point out all the places on their body that have been plagued by cancer without at the very least feeling heartsick.

Indigenous downwinders who raised families around the uranium industry recount clusters of family illness with a particular element of anguish. While those affected by the test site consistently note they would have moved their families away if they had known their homes were in a dangerous area, indigenous peoples affected by the uranium industry often relate how they moved their families closer to the site of contamination in an effort to reunite with their loved ones working in the mines or mills. As with other downwind residents, they were ignorant of any danger at the time, but now they must live with both the compounded guilt of having potentially exacerbated

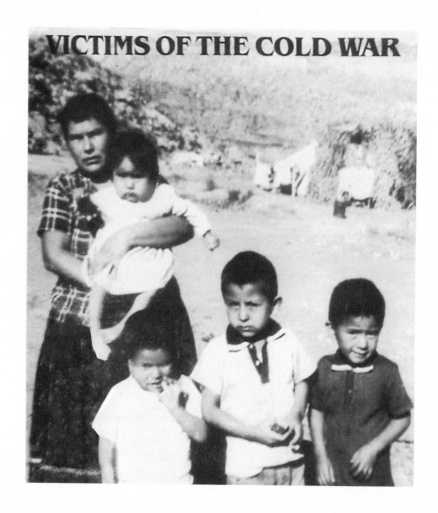

Fig. 8. "Victims of the Cold War," the Badoni family near a uranium site in Slick Rock, Colorado, 1960. This historic photo was used on a poster advertising the Native American Day in Remembrance of All Uranium Workers and Their Families Convention, 24–25 July 1999. Gilbert Badoni, the child standing on the far left, later testified that every member of his family had developed cancer or lung disease. Poster in the collection of Phil Harrison.

their families' exposures and the anger that comes with knowing their government stood back and watched them do it.

Lorraine Jack from the Shiprock, New Mexico, area, described how as a young wife she "would live there where [my husband] worked. . . . Pretty soon we had children. . . . As a pregnant mother I used to go there [to the mining camp]. So, the babies were born there." As a child, Jack had grown up near the Shiprock uranium operations where her father worked. In 1995, she told interviewer Phil Harrison how her children "played in the ore that was dangerous, just as we did way back. I did not know it was dangerous. So we exposed all of them to uranium, just as we were exposed to it before." Having learned about the toxicity of uranium, Jack reflected on her children's health problems. "One of our children who was eight years old, perhaps nine, that one was hospitalized. They told us his heart or lungs were not very strong. So, for that reason, he is under doctors' care and gets a shot every month from the pediatrician. The other one is our first born, who is now twenty-two or twenty-three years old. Back then when he was in junior high he used to say he had pain in the chest. He was told his heart was not very strong also. . . . So I wonder if it was from the uranium." This persistent wondering if any and all health problems can be tied to radiation exposure is a hallmark of life in the downwind and uranium-producing regions.

Navajo miner George Lapahie's children "used to ride their bikes on the tailing pile to play." As they aged, they developed what their father described as "tumor problems. What is it coming from? Through their investigation they have traced it to the uranium. One had serious work done on his head. His skull was cut and he had radiation treatment. That is how it is. Another was affected in their internal organs. My sons and daughters are like that." Echoing many downwinders and uranium-affected people, Lapahie pointedly noted, "In the past there were never stories about this. Now . . . our children are affected by it."29

Minnie Tsosie of Cove, Arizona, attributed her mother's and

husband's deaths, as well as her children's numerous health problems, to uranium exposure. "Today, three of my daughters are like that. They are told they have cancer inside of them. The cancer has affected them in their uterus . . . and they cannot have children." Tsosie's eldest daughter suffered from "something in her stomach, and they are thinking it to be uranium-affected. She has been asked if her father worked with it. 'This is what they suspect is affecting me,' she said." One of Tsosie's sons "experienced a problem with his nose. There was something that was affecting his breathing, that shut off the breathing. . . . The other son, his name is Lewis, was having a problem with his ear. There was a lot of pus that was running out of his ear when his father worked in the mine. Pretty soon the pus destroyed his eardrum, he was told." As a mother, Tsosie bore the weight of her children's ailments as well as her own. "We were all exposed to it. I am like that too. . . . I get the fever at night and I become fatigued. . . . I lose my appetite. I do not think of food. This occupies my thoughts. I think this is what uranium does, and we lived in it. It becomes overwhelming."[30]

Uranium-affected families and test site downwinders consistently note how those who worked, played, or spent time outdoors were disproportionately affected by radiological contamination, and they explain their contamination was amplified by their reliance on locally produced food. Glenna Orton began her narrative with a recollection of witnessing a test: "The cloud itself was beautiful. It just raised up like a big mushroom. I think it was Dirty Harry, the wind kind of stopped and it just kind of hung right over these west hills for three days." By describing the way wind and local landforms affected the movement of the cloud, Orton rooted her story in local knowledge. Invoking Dirty Harry added credence to her account for local audiences, drawing on the regional memory of the well-known contamination associated with the infamous 1953 test. Rare is the downwind narrative that does not reference Harry; some storytellers who are too young to remember 1953 confidently offer their memories of the Harry cloud. Rather than offering proof that the stories are local legends, these historically questionable Harry references only underscore the profound trauma of the atmospheric-testing era for downwind residents. Harry symbolized the betrayal of the era for them the same way Khrushchev came to symbolize the communist threat for Americans everywhere.

"They kept saying it won't hurt you," Orton continued. "They brought monitors to put on several people. When the Geiger counters started to go crazy some called and tried to get them to come and check, but they would not do it. Ross Hewlett, his was an immediate death and they did sign his death certificate as radiation poisoning. He has had one daughter die of cancer, and another who had a malignant kidney removed last year." Orton recalled how Shot Harry "left a salt film over the cars and everything was tinged a little pink. It was all over the sidewalks. It almost had a salty taste to me. They said, 'Wash your cars, the radiation is a bit high.' My father said, 'How do I wash my barns, my animals, farm equipment, and my fields?'" Orton described her family and neighbors as being "back to nature as much as a people could be. We all grew gardens, do [sic] canning, raised our own animals, we had milk, our own cows and chickens. We ate meat and fruit and things growing in the ground and we were being re-contaminated with it."

Orton watched her father, who ran sheep and cattle "up on the mountain by Brian Head," and her mother succumb to cancer. She tallied other local losses: "There were four good-size farms there. Out of those people, my father, my mother, and my brother had cancer, and Dad's hired man. Earl Bunn who had the other farm has had cancer for quite some time, the two brothers both had cancer, they lost their hired man too, and one man's wife had cancer. She's dead." Orton interrupted her accounting of human illnesses to tell Gallagher: "You should have seen the lambs. They were born like little balls, no wool on them, just deformed, pitiful little things. Randall Adams would never eat the mutton that he had in his herd because when he butchered them he said there was a black streak along their spines and he was afraid of them." Orton's narrative mixing of human ailments and livestock ailments illustrates vividly her understanding of the way nuclear contamination permeated the local food chain. "Less people would have been affected if they had let the cloud go over Las Vegas, because people lived indoors and worked indoors and they didn't raise their own food," she affirmed.

Evan Cooper of St. George, Utah, told Carole Gallagher a similar story. "About ten o'clock the clouds would begin to come through here and they weren't rain clouds. They were more like a sunset cloud, reddish, and they would fan out into the valley." Cooper explained that he did not think "there would be any harm done . . . until people began to die with leukemia, young children. We just had one case after another, and we began to wonder. There was five within a block that died." After describing his daughter's thyroid problems and the death of the mother of one of her schoolmates, he mused, "It seemed to hit the outdoors people more than the indoors people . . . fallout fell on the ground and we grew our own vegetables. We had our own chickens, pigs, beef, milk, and butter. We were sitting ducks because we made our living off the soil. It was a funny thing, you could go out with a Geiger counter anywhere and put it down and it would just go wild."[31]

Claudia Peterson remembered some days when her mother kept her children indoors: "We thought we were being punished. Because we lived out on a farm south of Cedar City and we played all day, every day, outside, and there was a couple times where she wouldn't hang clothes on the clothesline and didn't let us go out to play. And that was a very odd thing. Because we spent all day, every day, out playing." Claudia's mother kept "a huge garden, and we grew up around fruits and vegetables. . . . Our milk we got from the neighbors down the street." Peterson inadvertently witnessed a nuclear test while playing in the yard at the age of five or six. "I remember it because I was outside with my brother and I saw this big red ball come over up over the horizon and I thought it was a flying saucer [laughs] so I ran to the house to tell my mother. . . . When I got back out it had dissipated into a cloud. . . . I didn't know what it was. Put the connection together later on."

The connection Claudia made did not stop with an association between the "big red ball" and nuclear testing. In the 1980s, cancer began to sweep through her family. Her father died of a brain tumor. Her sister, Cathy, was diagnosed with melanoma, which Claudia found "very odd. Because I was the tomboy . . . we spent a lot of time on Cedar Mountain and outdoors, me more so than her." Cathy battled tumors in her brain, lungs, liver, bones, and breast tissue before passing away in 1987 at the age of thirty-six, leaving behind six children. A month later, Claudia's six-year-old daughter, Bethany, succumbed to acute monoblastic leukemia after an excruciating three-year battle.

Prior to her daughter's diagnosis, Claudia had not actively connected her family's losses to radiation exposure. She had heard rumors about illnesses being connected to the test site, but she described herself as "just naïve and dismissive. If it's not going to happen to you, you just blow it off." She remembered a doctor suggesting her father's tumor "was probably related to fallout exposure. But he said he couldn't prove it. . . . I wasn't an activist or anything at that point, although I started becoming more and more aware of it." After Claudia's father passed away, Bethany, "my little girl, was diagnosed with cancer. And I knew. I knew that the picture was diagnosed. I knew it. Because hers was a germ cell that's in embryo, that for some reason goes crazy in these kids. And I just knew it [was connected to radiation exposure]. The doctor said to me, 'Well, we don't know that.' He said, 'You couldn't probably prove that.' And I said, 'I don't need to.'"

Confronted with a knot of losses in her own family, Claudia began to connect the "flying saucer" from her childhood to "piles of dead lambs " and, in fourth grade, an adult waving a Geiger counter over her face at school. The meter "went bananas, and I said, 'What does that mean?' He said, 'It means you've had dental x-rays.' I knew that

I had never had a dental x-ray, my mother being a nurse." Two years after the Geiger counter incident, she remembered that "one of the little boys that was a year behind [in school] died of leukemia," and "in eighth grade one of my friends had his leg amputated from cancer and died a couple months later." These childhood memories, reexamined as her immediate family battled cancer, became the historical context of her personal narrative as a downwinder. Claudia's family history of radiation exposure is not uncommon in the downwind region for she also traces their losses to the uranium industry. Her husband's father, a uranium miner in Marysvale, died of cancer when her husband, Phillip, was only twelve years old.[32]

For years, Elmer Pickett of St. George, Utah, carried his list of names in his wallet. Interviewed for the 1994 ABC documentary *Cover-up at Ground Zero*, Pickett stated: "I have lost fourteen people, that's immediate family, [among them] my wife, my sister, a niece, my grandmother, my mother-in-law." Pickett described how his wife's "first symptoms were late in [nineteen] fifty-nine, we thought at first it was flu, then they discovered that she was full of lymphosarcoma, known as Hodgkin's disease, which is radiation induced." At the time of his wife's diagnosis, the Pickett family had six children. Elmer remembered his wife "desperately trying to train her real young daughters on how to take care of things. She just seemed like she was obsessed, almost, in getting those things done. And then, just before Christmas all at once, she come out with full-fledged leukemia. It took her within a week. She was thirty-eight."[33]

Raised in Salt Lake, Mary Dickson survived thyroid cancer in her twenties and became familiar with the plight of southern Utah downwinders as a peace activist and journalist in the 1980s. She never connected weapons testing to her own cancer until she met Carole Gallagher, who was in the region interviewing downwinders. Gallagher asked Dickson about her childhood milk consumption and showed her a map created by researcher Richard Miller that illustrated fallout deposition across the entire state of Utah, as well as

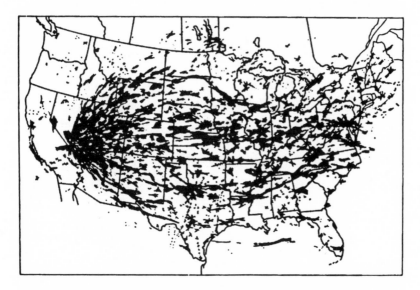

Fig. 9. An activist postcard, "Areas of the United States Crossed by Three or More Clouds from Atmospheric Testing, 1951–1962," featuring a map created by Richard Miller. Downwinder activists frequently reproduce this map to raise awareness of the reach of test-site pollution. From the collection of Mary Dickson.

most of its neighboring states to the east. "There's still a part of you that wants to think, 'It didn't happen to me,'" Dickson reflected in a 2005 interview, sitting at her dining room table. "But it did. And it keeps happening to people I know."[34]

While attending a hearing on government compensation of downwind residents, Dickson encountered a woman who had grown up nearby. "She came up to me with this yellow legal pad of paper, and she said, 'You know, I'm from your neighborhood.' And she shows me this list, and she says, 'These are all the people I know who got sick there.'" Dickson looked over the list and recognized the names. "They're all the kids I played with, and the kids I knew, so I called my mom. And I said, 'What about this?'" Dickson's mother had her own list of deaths in the neighborhood, as did Anne Dickson DeBirk, Mary's sister. The women began "merging [the] lists. This list just keeps growing and growing. And there were people being added to it all the time, and it gets up to like forty-five people, just in a five- or six-block area."

On 18 March 2001, Dickson's sister died after an eleven-year battle with lupus, one of a number of autoimmune disorders that occur frequently among downwind populations. Several years after her sister's passing, Mary memorialized Anne's struggle and explored her own journey into awareness as a downwind resident in her play *Exposed*, which opened to critical acclaim in Salt Lake City in October 2007. After the final scene, actors read aloud the list of deceased friends, family, and neighbors that Mary and Anne had begun compiling years earlier, a list that now numbered fifty-five, including Anne. Those attending the play were invited to add the names of those loved ones whose deaths they connected to the test site. Within three weeks, the list had nearly tripled.[35]

Ultimately, the list making, the connections, and the identified clusters are all supposition. Downwind residents realize they cannot prove that simply because a neighbor worked outdoors or a child drank milk from a particular dairy, their cancer was a product of exposure to radiation. Even the families of test site workers, atomic vets, and uranium workers, with their direct and irrefutable link to radiation exposure, know that no matter how dramatic the evidence seems, they cannot demonstrate conclusively a causal relationship between radiation exposure and a particular cancer death.[36]

For those who have tied their losses to radiation exposure, lingering questions of definitive proof are less significant than the weight of their tragedies and the knowledge their government deceived them about their exposure. Because no one can prove their illnesses are *not* related to radiation exposure, it is that possibility that is so powerful. When Bethany Peterson was diagnosed with leukemia, "the first thing I asked [the doctors] was, 'Is it from the radiation, is it from the nuclear testing we were exposed to?'" Claudia remembered. "I said, 'Could you prove it wasn't?' And he said no. I knew right then, I just knew, that that was what it was from. Because of what I was exposed to and what my husband was exposed to as children."[37]

Baby boomers from the downwind and uranium-producing regions consistently describe growing lists of deceased schoolmates at class reunions, epidemics of infertility problems among their peers, and a pervasive unease about the safety of their local landscape and food supply. Many describe a sense of foreboding upon turning to their local newspapers' obituary section, which they read regularly. Some make a habit of clipping those obituaries of friends, family, and strangers whose deaths they connect to radiation exposure, and they often bring out these clippings files to show journalists or interviewers. Downwinders often explain that if the deceased in a particular obituary is unknown to them, they read for the date of birth, cause of death, and the deceased's hometown, attempting to construct for themselves some indication of the person's likely radiation exposure. Each new death or diagnosis a downwinder or uranium-affected person perceives as connected to radiation becomes a part of their unique cosmology of loss. Deaths are registered geographically, in relation to the perceived avenue or avenues of exposure, and are linked to other clusters of similar disease and to clusters of loss in particular families, careers, generations, neighborhoods, and so on.

Memories of shared regional experiences during the atmospheric-testing era and the first uranium boom heightened the power of these stories as Irma Thomas, Claudia Peterson, and other individuals relayed them to others in their communities. Local newspaper coverage from the 1950s reveals that despite fears over the global spread of communism and the looming implications of the Cold War, downwind communities shared in a national sense of optimism. New technologies seemed to promise great things for the home, the farm, the community, and the nation.

Military facilities dotting the Great Basin, such as the NTS, Nellis Air Force Base, Indian Springs Air Force Base, Hill Air Force Base, Tooele Army Depot, and Dugway Proving Ground, provided substantial economic boosts to small agricultural regions, creating jobs and increasing population and local infrastructure, just as the uranium industry did elsewhere in the region. A generation of young men home after military service in World War II began families and stepped into leadership roles in their communities' farms and businesses. Veteran Sheldon Johnson, who fought in the Philippines, opened an optometry practice in St. George and joined the chamber of commerce upon his return from the war. Quentin Nisson, a survivor of the Battle of the Bulge, purchased a small store across from the elementary school in Washington City, Utah, and shortly thereafter was elected mayor, a post he held from 1950 to 1964. Many of their peers attended college on the GI Bill and did not return to agricultural work. The chambers of commerce in St. George, Cedar City, and other communities began to develop industry and tourism to provide capital and employment opportunities for the growth of their towns. Taking advantage of the rise in automobile tourism, many southern Utah communities began to advertise themselves as gateways to nearby national parks and as unique living spaces to draw tourists and new residents. Cedar City's annual Shakespeare Festival, today a fundamental part of the community's identity, began as part of this effort.[38]

The postwar baby boom meant scores of women in the downwind region carried pregnancies and gave birth during the 1950s, in the thick of the atmospheric-testing era and the initial uranium boom. The Iron County (Cedar City UT) Record ran a half-dozen birth announcements and as many one-year birthday greetings for local children every week, and the number of children in the downwind region rose steadily throughout the 1950s and 1960s. The small newspapers of this era frequently ran birth and birthday announcements on the front page. Occasionally they adjoined news of the test site in an eerie foreshadowing of the infant mortalities, birth defects, and childhood illnesses that would begin to spread through the region, presumably the result of the mothers' exposure to radiation while pregnant.[39]

Local newspaper coverage in this era—the *Iron County Record*, published out of Cedar City, and the *Washington County News*, published out of St. George—consistently focused on local growth but noted only occasionally the activities at the test site, a short distance west. The *Deseret News*, based in Salt Lake City, reported on NTS activities with much greater frequency and was at the forefront of journalistic efforts in the 1970s to expose the impact of testing on Utah residents' health. During this era of exposés, southern Utah papers remained relatively timid in discussing the issue of radiation exposure. Why was the more widespread local outcry absent in the local papers, when people elsewhere in the nation were publicly raising questions of the safety of testing and specifically the radiological contamination of milk?[40]

While this question is difficult to answer conclusively, it is of vital historical importance. Residents of the downwind region who do *not* recall unusual events or illnesses have pointed to the lack of local newspaper coverage of those events as evidence that nothing out of the ordinary ever occurred. In October 2005, retired physics professor Daniel W. Miles of Washington City published "The Great Fallout-Cancer Story of 1978 and Its Aftermath," an article discrediting downwinder stories. His article drew on some scientific reports, but he relied heavily—as did the downwinder stories—on personal memories of life in Washington County during atmospheric testing and on conversations with his peers. "I have read all those weekly columns found in the *Washington County News* during the weapon testing period," Miles claimed. "Nothing, absolutely nothing, is found . . . about people afflicted with hair loss, badly burned skin. . . . Nothing about towns covered with fallout 'snow' or vehicles needing a new paint job."[41]

His recollections and those of like-minded individuals are just as valid as the memories of those citizens who do recall such incidents; however, citing the lack of newspaper coverage does not constitute proof that these events failed to occur. People of that generation

A-Bomb Blast Lights Up Western Skies

Residents who were up and around Monday morning at 6:20 saw the western skies light up as the detonation of the long awaited Granddaddy A bomb at Nevada Test site was exploded by the Atomic Energy Commission. This view taken at the time of blast gives vivid outline of buildings housed in darkness just seconds before. Center of blast was just over the top of building in center of picture. The western horizon was ablaze for about 20 seconds.

Fig. 10. The *Iron County (Cedar City UT) Record* prints report "A-Bomb Blast Lights Up Western Skies," news of a "Granddaddy A bomb" test as part of the Operation Teapot series, next to the day's birth announcements on the front page, 10 March 1955.

tended to keep suffering relatively private. Obituaries during that era do not describe traumatic deaths in detail but refer to "accidents" and state individuals succumbed to "lingering" or "sudden" illnesses. Additionally, many leaders in communities immediately downwind of the NTS did not—and still do not—entertain talk of possible danger from the test site. After receiving personal reassurances from AEC officials working to promote the goals of the zone management program, many local politicians and business leaders became dedicated advocates for the test site. Dependent on the advertising monies and goodwill of these local leaders, it is unlikely that small local newspapers would have covered rumors of atomic danger or possible radiation-related illnesses. The Crandalls' *Tonopah (NV) Times-Bonanza* stands out as an exception.[42]

During the atmospheric-testing era, Mayor Nisson of Washington City contacted the mayor of neighboring St. George to ask his opinion on the possible danger that testing posed to their communities. Nisson was uneasy about the radiation monitors who were collecting

Nuclear Device Raises Tower At Yucca Flats

A nuclear device, the sixth in the current series at Yucca Flats in Nevada, of about medium strength was detonated Tuesday morning at 6:05, MST. Many residents of the southern Utah area were up and around when the brilliant glow appeared over the western horizon. In Cedar City residents heard a distinct rumbling, 14 minutes following the blast, but no sharp cracks as heard with the larger test.

According to Dr. Clinton Powell, senior member of the Atomic Energy monitoring team located in Cedar City, no measureable fallout was recorded in this area. Reports from the Las Vegas Test information office indicated that the cloud from the explosion rose to about 40,000 feet before being blown slowly toward the ESE with wind speeds of five miles per hour near the surface and as high as 53 miles per hour.

Fallout was recorded in Las Vegas, Nellis Air Base, Henderson and Boulder City.

After the detonation observors in the Control Point area saw melting pieces of the tower falling from the rising column. The entire tower, observors report, was shattered, leaving only tower stubs and the concrete anchors of supporting guy wires.

Fallout on Las Vegas and vicinity was also very low and without any effects on health. In North Las Vegas the total was 0.174 roentgen in estimated biological exposure for one year, compared to an estimated guide of 3.9 roentgen per year by test organization officials.

BIRTHS

March 16 through March 23

A girl, born to Mr. and Mrs. James Sandberg, Cedar City, Mar. 16.

A girl, born to Mr. and Mrs. Garth B. Nelson, Cedar City, Mar. 16.

A girl, born to Mr. and Mrs. Elmer Barrick, Cedar City, Mar. 17.

A boy, born to Mr. and Mrs. Edwin Gale, Cedar City, Mar. 17.

A boy, born to Mr. and Mrs. Wayne Holt, Beryl, Mar. 17.

A boy, born to Mr. and Mrs. Clair Carter, Cedar City, Mar. 17.

A girl, born to Mr. and Mrs. Delmont Wallace, Toquerville, Mar. 17.

A girl, born to Mr. and Mrs. Clark Barton, Cedar City, Mar. 19.

A girl, born to Mr. and Mrs. Richard Urie, Cedar City, Mar. 20.

A girl, born to Mr. and Mrs. Paul Wood, Cedar City, Mar. 20.

A boy, born to Mr. and Mrs. Ross Sargent, Cedar City, Mar. 20.

A girl, born to Mr. and Mrs. Wilson Fritz, Cedar City, Mar. 20.

A girl, born to Mr. and Mrs. Robert Heyborne, Cedar City, Mar. 20.

A girl, born to Mr. and Mrs. Rodney DeMille, La Verkin, Mar. 21.

A girl, born to Mr. and Mrs. Hal Heyborne, Cedar City, Mar. 23.

Fig. 11. Test site news from the Operation Teapot series printed adjacent to the birth announcements on the front page of the *Iron County (Cedar City UT) Record*, 24 March 1955. The account states that "no measurable fallout was recorded in this area."

data at the local schools. He remembered, "I asked the principal over here if they ever give any report of what they're finding about the radiation in the atmosphere, and he said, 'Nope, they just say that this goes in to be reported, and we have nothing to tell you.' I said [to Clint Snow], 'It makes me kind of uneasy,' you know?" Mayor Snow did not share Nisson's concerns, or if he did, he was reluctant to admit it, a fact Nisson attributed to the economic boost the local military activity was giving St. George's economy. "His attitude wasn't very good," Nisson recalled, surmising Snow's reluctance to criticize the government was "because of this business in St. George. They had the government in there with planes, maneuvering around. I don't want to be too critical of him, just the fact that he was kind of easygoing with some people. 'Oh, my, we've got to be patriotic, the Russians is throwing a Spudnut [*Sputnik*] up,' you know, so they got off pretty slick there for a while."

Irma Thomas's husband, Hyrum Thomas, served as president of the St. George Chamber of Commerce in the 1950s and was an active local leader for subsequent decades. Their daughter Michelle described how her father "would get some companies interested in coming [to St. George], and then it was his job to get them in. And then they'd take them out to dinner and talk, and the women would say, 'Well, tell us about the schools.' And my mom would launch into her 'don't bring your children here, this is the most toxic place on earth.'" Irma's honesty won her few friends among St. George's boosters. "The city fathers were very angry at her," her daughter recalled, "because how was that going to help attract people here if she ran off at the mouth?" The city fathers weren't the only ones angry at Irma. Her vocal opposition to weapons testing and her willingness to criticize male civic, military, religious, and government leaders who supported it were an anathema to many of her LDS neighbors and female acquaintances. "She took so many cold shots," Michelle remembered. "They were so mean to her, but I never saw her cry. She was on a mission. It didn't matter what people thought, if they called her a troublemaker or thought she was unpatriotic. She was single-minded in her purpose: to protect her family."[43]

In 1995, while researching an article on downwinders in St. George, reporter Ken Scharnberg "was informed by [a] businessman that the radiation question had been kept alive by a handful of malcontents, and was 'bad for business.'" Claudia Peterson noted in 2005 that at times she has had to "walk a fine line" in her efforts to bring attention to radiation-related illness because of her husband's position as the city finance director.[44]

Activists trying to bring attention to the risks posed by the uranium industry faced similar resistance in the 1970s. Lured by the prospect of economic solvency and industry promises of new safety standards, laborers wanted nothing to do with Manny Pino's and other activists' warnings. Pino recalled how their early efforts "to create public information, public awareness, to both the mining and non-mining population in this area of the Laguna Pueblo" were received. "When we would take documentation about how the body ionizes uranium, and other nuclear contaminants, and [ask] the workers if they had been given medical screening, most of the time the answer was no, and on certain occasions the mining population, the workers didn't want to talk to us. They'd say, 'Get that shit the hell out of my face.'" Representatives of the tribal government asked Pino and other activists, "Why are you trying to scare the people? They're only trying to make a living. These are good wages!" He described the prevailing mind-set as "complete denial."

Most community leaders willingly cooperated with AEC and uranium company officials to reassure their towns and neighbors. Without access to other expert opinions, these leaders relied on the government or industry representatives' assurances of safety. Additionally, as Quentin Nisson suggests, leaders such as Mayor Snow were likely disinclined to appear unpatriotic, cause alarm, or sacrifice economic growth in their communities by heeding rumors of danger. Individuals who attempted to draw awareness to the apparent risks of radiation exposure engaged in an uphill struggle. Despite the difficulty they encountered in persuading others to heed their stories, the early list makers continued writing down names, gathering information, and grounding their stories in the landscapes, the lifestyles, and the local histories of their communities. They found other list makers, as Irma Thomas and Mary Dickson did, and they combined their lists, which grew longer with the years. Initially few thought of themselves as activists, but they were laying the foundation of a movement.

{6} CRITICAL MASS

Could it be that other bureaucrats are sweeping other things under the rug now as the AEC did in the 1950s?

Deseret (Salt Lake City) News editorial, February 1979[1]

In the domain of nuclear physics *critical mass* is defined as the smallest amount of an elemental material that can set off and sustain a chain reaction, given the right conditions. Multiple factors play into the achievement of critical mass, ranging from the temperature or shape of the material to the degree it has been enriched or altered. For some elements, only a small amount is needed to achieve a nuclear chain reaction. Other elements must be present in larger quantities, or must be profoundly affected by temperature or pressure, before they will sustain a nuclear chain reaction. Once critical mass has been achieved, a tiny amount of extra material can cause the reaction to go supercritical, vastly accelerating the speed of the chain reaction.

Ironically enough, the critical mass of nuclear material is an excellent metaphor for the critical mass of awareness that sometimes occurs among groups of people. We can define this *social critical mass* as the smallest number of people who are cognizant of a particular problem and who, under the right social and cultural circumstances, can facilitate the continual exchange of information about that problem.

Individuals such as Irma Thomas who became convinced of the dangers of radiation exposure early on narrated their observations to others in their social circles. While many listeners discounted these observations, a few were convinced when the story they heard resonated with their own experiences. Some of them then began to narrate their own observations about the problem in their particular circles, and awareness of the problem rippled outward. As these persons repeated their observations to new audiences, they created their own version of a story about life downwind of the test site or near the uranium industry. They told their stories to make sense of their losses, to bear witness to the suffering they observed, and to convince others of the risks at hand. Each retelling further established the story's structure and the storyteller's delivery and narrative style. Retellings also served to strengthen their belief in the contamination and betrayal they described. When enough people found the story compelling that it continued to be retold, social critical mass was achieved.

The ability of storytellers to generate a critical mass of awareness regarding a problem in their community depends on how effectively they narrate the problem to those they wish to convince. Listeners more easily believe storytellers who ground their narratives in specifics and draw on shared local knowledge to make their case. They referenced events that loomed large in community memory, such as the sheep die-off in Cedar City, the advice to take cover in St. George after the Harry test, or the deaths of the fathers in Red Valley. Narrators also alluded to phenomena generally known about but not always connected to radiation, such as the women's miscarriages or the uranium workers' breathing problems. Using regional details, such as place names and the names of local families, demonstrated the storytellers' roots in the region and proved a crucial step when trying to convince patriotic neighbors that they had been betrayed by their government and that the safety of their local food, water, and air was now suspect.

It was difficult in the 1970s for even the most talented storyteller to convince a majority in his or her community that nearby nuclear tests or uranium sites had wreaked damage, and it remains challenging today. Accepting the possibility that invisible contamination has seeped into the local environment and the bodies of loved ones is terrifying, and many people found (and still find) it easier to dismiss such claims as conspiracy theories promulgated by crackpots, a dismissal allayed by the credentials and reassurances of government and industry authorities who claim radiation never exceeded "safe levels."

To achieve critical mass, though, early activist-storytellers such as Irma Thomas and Manny Pino did not need to convince everyone in town, as much as they would have liked to do so. They only needed enough people to start paying attention for the storytelling to achieve a self-sustaining momentum. Once that momentum was established, activist-storytellers could push for *supercriticality*, or a rapidly increasing level of awareness of domestic nuclear danger that, when combined with particular events and cultural factors, could fuel a shift in regional and national attitudes. This change could force the government to admit it had been less than forthright about the dangers inherent in domestic nuclear development, and activists could use that acknowledgment to push for policy changes, toxic site cleanup, and assistance and remediation for those whose health and families had been affected.

Supercriticality did occur in the late 1970s, as persistent storytelling converged with several events on the national stage. The country's long and tragic military engagement in Vietnam, coupled with the Watergate scandal, had severely undermined the integrity of the monolith that the American people saw as "The Government." Nearly two decades after the publication of Rachel Carson's *Silent Spring*, the nascent environmental movement had succeeded in familiarizing many Americans with the concepts of toxic pollution and the contamination of ecosystems. Whistleblowers, such as Kerr-McGee employee Karen Silkwood, began calling attention to the lax controls on safety and contamination at many of the nation's nuclear facilities, and Silkwood's suspicious death in a 1974 car accident only fueled public speculation about corruption and cover-ups in the nuclear industry. Journalists with a taste for muckraking found receptive audiences for pieces dedicated to exposés of government and industry corruption. The energy crisis precipitated by the Organization of Petroleum Exporting Countries' 1973 oil embargo made nuclear power a popular topic, and nuclear power plant construction began in many communities around the United States, revitalizing the temporarily defunct uranium industry. This convergence of circumstances set the stage for the debut of downwinder and uranium-affected storytellers on larger regional and national stages.

These storytellers rarely resembled the stereotypical activists American society had come to associate with the political movements of the late 1960s, nuclear disarmament included. They were not hippies or students, nor did they hail from the more progressive coasts. They were heartland residents: mothers and housewives, ranchers, war veterans, and blue-collar Americans. Most consistently stressed their patriotism, humility, and rootedness in the American landscape. Many even went so far as to argue for the importance of the nation's nuclear arsenal, emphasizing they only wanted that arsenal built safely and with careful attention paid to the release of radiological contaminants. As a result, downwinder and uranium-affected narrators could successfully appeal to more conservative American audiences, who often felt disinclined to support other progressive movements promoted by more radical-looking and sounding activists.

Regionally the downwinders broke into the broader public

consciousness on 12 August 1977, when readers of the *Deseret (Salt Lake City) News* learned "a farmer working in a field in southern Utah may have the seeds of cancer in his veins, a result of radioactive fallout which drifted across the region many years ago." In his article "Deaths High in Utah Fallout Area," journalist Gordon Eliot White cited a "nearly forgotten" National Cancer Institute study, completed seven years prior, that indicated leukemia rates in several Utah counties "were nearly twice the state and national average." Though the article failed to merit the front page, it proved to be the opening gust in a storm of outrage and publicity during which the word "downwinder"—and the individuals that the word represented—became a part of the media and public lexicon.[2]

Within two weeks of White's article, Utah's new Democratic governor Scott Matheson began to push for a full review of previously classified Atomic Energy Commission documents from the 1950s and appointed an inquiry commission to pore over the documents when they were obtained the following year. Matheson's personal story informed his gubernatorial actions: as a southern Utah resident, hailing from Cedar City and Parowan, he had witnessed "brilliant flashes and thunder-like blasts" from nuclear tests in the early 1950s, and many in his family struggled with health problems afterward. Matheson found the stories of other downwind residents compelling and convincing. He became one of their most powerful advocates, speaking and acting forcefully on behalf of downwind Utahans throughout his two terms as governor. In the mid-1980s, Matheson was diagnosed with terminal multiple myeloma, which he connected directly to his radiation exposure from the Nevada Test Site. In 1986, four years before he succumbed to his disease, Matheson stated that the downwinder stories "point[ed] to a continuing need for governors to be vigilant concerning both short-term and long-term impacts of federal decisions on their residents. If citizens in a state are to be sacrificed for the 'national interest,' then, at the very least, those citizens need to be fully informed and protected as much as possible."[3]

The public clamor for answers about radiological contamination that followed White's 1977 article and Matheson's aggressive demands for information almost immediately precipitated a formal study of leukemia in southern Utah. University of Utah medical doctor and scientist Joseph Lyon headed the effort. Within months, class action lawsuits brought by those people living downwind of the test site and the uranium industry began to coalesce.

Journalist Gordon White and Governor Matheson did not stumble inadvertently on the issue of radiological contamination; Irma Thomas and other activists had been demanding they investigate it for years. Thomas penned repeated letters to journalists, politicians, and government officials, letters her daughter described as "vituperative little missiles." One of those letters bore fruit: an anonymous government official sent Thomas a package of classified records relating to the test site. The Thomas family referred to the sender as a "deep throat," after the secret informant who tipped off journalist Bob Woodward to the Watergate scandal. "Mom never told us who it was," Michelle related later. "But she got the declassified [documents] long before Hazel O'Leary [secretary of energy under President Bill Clinton] let them out." The documents referred to the residents downwind of the test site as a "low-use segment" of the population and identified the Thomas family's hometown of St. George as "Fallout City." Irma Thomas never made the documents public. Her daughter explained, "The deep throat guy had told her, 'You can't publish this, because not that many people have access to it.' But she would reference [the documents] . . . it only fueled her more, and she wasn't about to give up."

By 1978 Irma Thomas was speaking to reporters from all over the country—and soon the international media—on a regular basis. A 10 July Deseret (Salt Lake City) News article profiled her work under the headline "Woman Fights for Safer Radiation Controls." It described her as "a light-hearted woman who laughs a lot about herself and the foibles of the world. One subject, however, is guaranteed to make sparks fly from this grandmother's eyes and that is radiation fall-out." The article quoted one of her letters to President Jimmy Carter: "What a runaround! What a passing of the buck! What do you think the AEC, the DOE [Department of Energy], or the XYZ cares about our health and welfare so long as we are submissive little guinea pigs? Are our lives of so little value that we must be sacrificed so your jobs can be carried on?"[4]

Irma Thomas was not the only mother fighting to bring attention to the issue of radiation exposure. Across the region, the widows of uranium workers were struggling to raise their children without fathers. Minnie Tsosie's husband "had just become middle-aged" when he succumbed to cancer. "Hardship is encountered when your husband's life ends like that," Tsosie related to Timothy Benally in 1995. "It is hard, no one to depend on; even though you have sons, they do not think of the home the way their father thought of the home." Tsosie summed up the experience of many mothers in the region when she reflected: "You get really tired, because you end up thinking of all these things yourself. It drains you. You do not want food. . . . You cannot sleep well. You just cry hard and feel better. And then you take a big breath again and start moving for your children." For some Navajo widows, moving again meant more than taking care of the home; like Irma Thomas, they saw a pattern to their losses and thought others needed to see it too.[5]

The uranium widows undertook "steep learning curves about science, politics, and organizing," in Doug Brugge and Rob Goble's words. Many were fluent only in Navajo and could not speak, read, or write English; further, most had no personal vehicles or telephones, creating significant barriers to networking and gathering new information about the disturbing losses in their midst. They persisted, gaining the assistance of local tribal leader Harry Tome. Their efforts garnered some press attention in 1973, and they persuaded some New Mexico State legislators to sponsor legislation on their behalf, but aid failed to materialize on the reservations. In 1978, Tome contacted attorney Stewart Udall, the former secretary of the interior under Presidents John Kennedy and Lyndon Johnson. Daunted by the prospect of approaching an individual of Udall's stature, Tome recalled that he was "tempted to go home and forget about it. I was just a little guy and a Navajo. But my people needed help."[6]

To his surprise, Tome found Udall receptive to his plea for assistance. Both men hailed from the same region of the Southwest, and both had deep roots and families in areas heavily impacted by nuclear development and the uranium industry. Udall was already in contact with Irma Thomas and had urged her to begin organizing downwind residents for a lawsuit. When Tome visited Udall, he offered the lawyer more than a crisis; he shared a story that resonated with his audience, a story that echoed others Udall had been told, a story that depicted a people Udall cared for deeply, a story that was rooted in a place Udall knew well.

As secretary of the interior from 1961 to 1969, Udall oversaw the policies of the Bureau of Indian Affairs during the later years of the first uranium boom and during much of the era when workers and their families began to fall ill. He inherited a BIA that for much of the previous two decades had promoted a policy known as termination. Designed to answer the clamor from political and business leaders to end the protected status of reservation land, termination policy sought to end the sovereign status of Native tribes. Using rhetoric about liberating Native people from the patriarchal oversight of the federal government, Senators Clinton Anderson (D-NM) and Henry "Scoop" Jackson (D-WA) argued the services and protections of the BIA were no longer necessary and that, for their own benefit, Native Americans should assimilate speedily into the national capitalist economy. Business interests stood to benefit mightily from this arrangement, as the mineral and timber resources previously held in trust by the BIA would be open for sale. Several tribes deemed ready for self-governance were "terminated," with disastrous results for those Native peoples and their lands.[7]

The Utah Southern Paiutes, who lived immediately downwind of the Nevada Test Site, were terminated in 1958, at which time they lost access to medical care from the Indian Health Service. A twenty-two-year gap in medical care ensued, during which time a plethora of health problems emerged, leaving termination and the downwind experience inextricably connected in the minds of many Southern Paiutes. Geneal Anderson, a former Utah Paiute chairperson, told Carole Gallagher her story in 1991. "I was a kid, born in 1952. You never were told what the pink clouds were. If you're not told, you don't pick up on what's happening." Anderson remembered "a little boy in the '60s, a cousin who used to come visit us all the time, one summer he came up and he was having a lot of nosebleeds. When they went back home to the reservation they found out he had leukemia. That was my first contact with it, in 1963 or 1964. He was four or five." In 1980 Anderson's tribe "was reinstated as a federally recognized tribe. There were a lot of health needs that needed to be met." Anderson's mother succumbed to leukemia 1986. "There was a drastic loss of weight, and she was bruising very easily. . . . She lived over near the Utah-Nevada border near the Indian Peak area when she was a child, and then [in] Cedar City."[8]

When the Southern Paiutes were terminated, Udall was a freshman representative from Arizona. He was beginning to take an interest in Indian policy issues, sparked in part by his time on the Joint Committee on Navajo-Hopi Indian Administration during 1957 and 1958. By the time Kennedy invited Udall to head up the Department of the Interior and, by extension, the BIA, termination policy had grown unpopular (as had atmospheric nuclear testing). Native self-determination, rather than termination of Native sovereignty, became the stated goal of Udall's Bureau of Indian Affairs. The underlying federal Indian policy goal remained largely unchanged as it sought to assimilate the reservations into the national economy, using their labor and natural resources as capital. In a 2003 interview Udall recalled, "We were always saying to the Indians, if we could just find a little economic niche for you, where you, because of your land and so on, could do something . . ."9

To that end, Udall's BIA administrators negotiated numerous contracts and land and mineral leases with various industries, including mining corporations, frequently without regard to the opinions of tribal officials or without stipulating adequate safeguards for environmental or labor protection. As noted in chapter 2, mining leases were often established for extended periods, without any provisions allowing for Native leaders to intervene, and set inferior rates for purchasing the minerals and labor. Such was the case when Udall's BIA set up a 1966 lease for the Peabody Coal Company to strip mine coal from Black Mesa, a landform in Arizona with local significance as a sacred site that touches both the Navajo and Hopi reservations. Believing the lease would provide Native people with substantial economic benefits, Udall supported it enthusiastically. Historian Thomas Clarkin notes the tribe had little opportunity to discuss or research the proposal; having been "told fossil fuel [soon] wouldn't be worth a tinker's damn" because of the advent of nuclear power, the Navajos approved it in a rush. Over the next thirty-five years, Peabody drained the local water table and racked up profits of $750 million, with no obligations for restoration aside from reseeding mined areas. Navajo royalties over that same period amounted to less than $60 million.[10]

Significant distrust of bureau officials among the Navajos led to frequent disputes between the tribe and Udall's BIA. Frustrated that Navajo tribal attorney Norman Littell did not share his vision for the Navajo people, Udall had Littell removed from his position in 1963. Littell filed suit. U.S. District Court judge John Sirica concluded that "the power to remove Littell resided exclusively in the tribal council" and condemned Udall's actions in 1965. Sirica's decision was later reversed in the U.S. Court of Appeals, and the rift between Udall's administration and the Navajos continued to widen. In a letter to the Johnson administration in the mid-1960s, Navajo Tribal Council member Annie Wauneka wrote that President Johnson "has made it clear that he will not tolerate corruption in his government . . . what then, does he intend to do about Udall?"[11]

Despite the frustration of tribal leaders, Udall continued to see himself as an advocate for Native Americans even long after his tenure as secretary of the interior had ended. When Harry Tome found him in 1978, Udall was ready to listen and ready to act. In January 1979 the former secretary drove to the Red Valley Chapter House to interview the surviving uranium workers and the dead miners' widows.

Udall described the Arizona trip years later in his book *The Myths of August: A Personal Exploration of Our Tragic Cold War Affair with the Atom*: "With the help of interpreters . . . I learned more than [I] ever wanted to know about lung cancer and about the epidemic that had settled like a plague over the families who lived in the vicinity of Cove and Red Rock and Lukachukai." Udall described the families as "pastoral, mostly illiterate Navajos" with a "childlike understanding of modern American life" that he attributed to the failures of the BIA educational system in the 1920s and 1930s. He was particularly moved by the widows, none of whom "knew about or had received workmen's compensation benefits for the deaths of their husbands." Udall was deeply affected by their stories, writing, "There were many memorable matriarchs (such as Grace Tuni, Rose John, and Rose Benally) who held their families together and somehow managed to give most of their children a high school education. When lung cancer claimed Rose John's husband, Tom, at age thirty-three and left her with five children ranging from six months to ten years, she did not falter. I inquired how she coped with her loss, and even got two of her children into colleges. Rose's reply was simple: 'Our family was strong. We always worked together and helped each other.'" The widows' resolve resonated with Udall, a devoted family man descended from generations of Mormon settlers who cherished community and hard work in the face of adversity.[12]

Udall's horror at the crisis the widows were encountering is clear in his writing. A shadow of paternalism that hearkens back to his days overseeing the Bureau of Indian Affairs is subtler but still persistent. As with every reformer in American history who has attempted to better the condition of indigenous Americans, Udall always believed he had the best interests of Native people at heart even if those Native people at the time disagreed. While a lifetime spent acting on government policies about what was "best for Indians" informed Udall's perspective, his description of the Rose Tom family indicates that later in life he developed more complex relationships with Native peoples, moving him to act as a fellow citizen informed by story and human interaction rather than by policy briefs.

These relationships moved Udall to launch a ten-year pro bono effort to bring compensation, health care, and some measure of government accountability to the uranium families. His wife and children became his assistants and fund-raisers. Their work with uranium activists had particular urgency, as the cycle of tragedy sparked by the uranium industry was beginning anew across the Southwest and elsewhere. Shaken by the hardship generated by the 1973 oil embargo and again by the 1979 energy crisis, Americans were evaluating their dependence on foreign oil. Increasingly, nuclear power was depicted as offering freedom from this dependence. The nuclear power industry expanded dramatically as a result, provoking a second uranium boom in the early 1970s.[13]

Ample scientific evidence already existed to substantiate a connection between the uranium industry and health problems in nearby communities. In July 1965, Joseph Wagoner and his colleagues published an article in the *New England Journal of Medicine* reporting on their ongoing study of lung cancer among uranium miners. Two years later Secretary of Labor Willard Wirtz testified before a subcommittee of the Joint Committee on Atomic Energy, arguing for increased protection of workers in the uranium industry. Wirtz pointed out that at least 98 uranium miners had died of lung cancer, and "another 250 to 1,000 — the estimates vary — [were] already incurably afflicted with it."[14]

In 1979, shortly after Udall's visit, journalist Molly Ivins visited Red Valley. The *New York Times* published her piece "Uranium Mines Leaving Indians a Legacy of Death" for a national audience in May. In her article, Ivins reviewed the troubling data about lung cancer continuing to emerge from the Public Health Service study and printed the response of officials at mining company Kerr-McGee: "We doubt that any evidence to support this allegation exists anywhere." Ivins also quoted LaVerne Husen, a doctor at the Indian Health Service hospital in Shiprock: "It got so I didn't need to wait for the tests. They would come in—the wife and the six kids were always there too—and they would say they had been spitting up blood. I would ask and they would answer that they had been in the mines. Then I didn't need to wait to make the diagnosis, really. I already knew. Those were the cancer cases." The month after Ivins's piece appeared, journalist Chris Shuey published his own article on the plight of Navajo uranium workers in the *Scottsdale (AZ) Progress Saturday Magazine*, thus focusing regional attention on the problem.[15]

These studies and articles received only minor attention in the communities where they mattered most—those adjacent to the uranium industry, where mining companies were still finding a ready workforce. The initial uranium boom in the 1940s and 1950s had fundamentally changed local economies, making many families completely dependent on the whims of a disconnected market. When uranium was booming, workers' families used their new incomes to upgrade their standards of living, and traditional practices of raising food locally and handcrafting clothing, arts, and tools had subsided in many areas. Decades later, Alveno Waconda reflected, "As I look back on my early teenage years, it angers me now to see how quickly money can change your whole life." He told his audience at the 1992 World Uranium Hearings how he and his siblings "began to have a different attitude about money and what it could provide. Those things which were a vital part of our childhood no longer seemed important." Once the AEC felt it had procured enough domestic uranium to sustain future weapons development and testing, the bottom of the uranium economy dropped out, throwing entire communities out of work. Those people who had stopped growing their own food in order to labor in the uranium industry found themselves in a dire predicament in the late 1960s, a plight made even more tragic and difficult for many when the primary breadwinners who had worked the mines and mills succumbed to illnesses and died. The jobs offered by the reinvigorated uranium industry proved irresistible to a new generation of laborers.

As it had during the first uranium boom, the proliferation of new jobs brought a wave of cultural changes. Manny Pino began studying and advocating for the Laguna Pueblo community of Paguate during the uranium boom of the 1970s and continued to do so for three subsequent decades. In Paguate, home of the massive open-pit Jackpile Mine, Pino witnessed "tradition and culture going to the wayside, because these people prioritized their nine-to-five jobs." He observed "a dwindling of participation in ceremonies in the traditional calendar that exists in these traditional villages. We began to see the language dwindling, because if you have a non-Indian speaking workforce present in close proximity to the reservation, you know, workers began communicating primarily in English." Local food production decreased as well. As he pointed out, "This used to be primary grazing land in the Laguna Pueblo. There were orchards, there were wheat fields, there were barley fields, alfalfa fields. . . . Everybody deprioritized agriculture, deprioritized livestock raising, right? Because thirteen dollars an hour working in the mine was a lot of money back then." High wages offered by mining companies proved a great temptation. "The dropout rate skyrocketed with ore production, because a lot of these young kids, seventeen, eighteen, even sixteen, found it economically more lucrative to work in the uranium mines than it was to sit in a classroom and earn a diploma."

The boom-and-bust cycle of the mining economy brought large numbers of outsiders into small Native communities like Paguate. As

Dorothy Nelkin describes, "Mining tends to draw on a mobile force of drillers, miners, heavy equipment operators, and engineers who follow jobs around the country. They compete for local services and form their own subculture, isolated from and often antagonistic to local Indians." New wages also brought dangerous new substances and social problems. Pino described changes in the town of Grants, New Mexico, "which prided itself as the uranium capitol of the world" and had a population of around eight thousand in 1950. "By the early nineteen seventies it was twenty thousand people. And you correlate wage earning with social problems. The jails were always filled to capacity; there was drugs in Grants, cocaine." Pino remembered cocaine use was rare in the region in prior years. "Nobody had the money to buy cocaine, you know? But by the late seventies, you could go on the res anywhere and buy anything: coke, acid, weed. Prostitution! In the city of Grants, that was unheard of before." In a literal illustration of the toxic effect of the mining boom, Pino described how the jump in population filled the "sewage lagoon in the city . . . beyond capacity. Where did it flow? Into the Rio San Jose."[16]

As the second uranium boom spread its toxic seeds, those planted by the first boom twenty years earlier began to come to maturity. Pino recalls, "By the early to mid-1980s, we began to see not only cancer clusters developing in the population but also children suffering from birth defects, women experiencing spontaneous abortions, Down syndrome. And the elders said, you know, 'These kinds of things just didn't exist among our people prior to this mining experience!'" Lured by the prospect of economic solvency and industry promises of new safety standards, however, laborers wanted nothing to do with the warnings Pino and other activists were offering.

Meanwhile, the media continued to publish revelations about health problems connected to radiological exposure. On 8 January 1979, the front page of the *Deseret (Salt Lake City) News* reported that a study from the 1960s demonstrating a "high cancer rate" in the downwind region had come to light, prompting Health, Education, and Welfare Secretary Joseph Califano to order a search of federal health files for other unpublicized cancer studies. The unearthed report, which federal officials claimed had been "misfiled," had been funded by the Public Health Service and directed by Dr. Edward Weiss. As early as 1964, Weiss had demonstrated "about a threefold excess of cancers among children who were under age nineteen living in . . . two southwestern [Utah] counties." A PHS thyroid study conducted concurrent with Weiss's leukemia research revealed no evidence of radiation-related ailments. Dr. Joseph Lyon later testified the thyroid study "was published, highlighted, and used to reassure the citizens of Utah [concerned about] the adverse effects. [Weiss's] leukemia study was buried in the files of HHS [Department of Health and Human Services] after a high-level meeting at the White House because of its impact."[17]

Childhood leukemias tend to appear within ten years of radiation exposure, while thyroid cancers take as much as twenty to thirty years to grow. Dr. John Gofman once observed that "radiation-induced cancer . . . add[ed] a new dimension to [the] problem of cause and effect." Because, in his words, "the time period between insult [radiological exposure] and disease can be measured in decades, not days, weeks, or months," assembling timely scientific data about radiation-related disease has always been problematic. Susan Dawson and Gary Madsen argue the traditional methods used to measure disease rates—cohort studies and case-control studies—"do not work as well when a lengthy latency period is involved." To maintain public support for weapons testing, meanwhile, the AEC delicately managed the mixed results of epidemiology studies. An early study showing no increase in thyroid disease was highlighted to reassure citizens they had nothing to fear from ongoing underground nuclear testing, while the concurrent Weiss study showing a leukemia spike went unpublicized. Fifteen years later, reports of the "misfiled" Weiss study created a storm of indignation in downwind communities and aroused the ire of Americans outside the immediate downwind region.[18]

Charged with "responding to newspaper accounts . . . that there was an excess of leukemia" in the downwind area, Dr. Lyon began studying the problem in 1977. He found the same heightened pattern of childhood cancer that Dr. Weiss had and later stated that he and his colleagues had "replicated" the Weiss study "without knowing we were doing so." Mere weeks after the revelation about the Weiss study surfaced, the Lyon leukemia study appeared in the *New England Journal of Medicine* and reported an increase in leukemia mortality of "2.44 times . . . greatest in 10- to 14-year-old children."[19]

The same month the Lyon leukemia study hit the papers, Governor Matheson's inquiry commission announced it had uncovered declassified documents that proved the government had lied to the Cedar City ranchers about the cause of the 1953 sheep die-off and lied again during the sheep ranchers' 1955 lawsuit. "Could it be," the *Deseret (Salt Lake City) News*' editorial department wondered, "that other bureaucrats are sweeping other things under the rug now as the AEC did in the 1950s?" In subsequent weeks articles discussing radiation, disease, and government cover-ups appeared in newspapers nearly every day and not just in Utah. National reportage on radiation-exposed citizens made the front pages of the *New York Times*, the *Washington Post*, and the *Wall Street Journal*. Congressional hearings were scheduled for April 1979 in multiple locations downwind; these field hearings would offer many downwinders their first opportunity to share their stories with the broader public and elected federal officials.[20]

Shortly before the hearings began, an accident in the nuclear power industry helped make "downwind" into a household word. On 28 March 1979, a partial-core meltdown occurred at the Three Mile Island Nuclear Generating Facility in Dauphin County, Pennsylvania. While many scientists later argued the incident did not create statistically significant disease trends downwind, at the time the incident caused a panic in the surrounding region and across the nation and spawned a new regional downwinder story trope. Less than two weeks before the accident, the movie *The China Syndrome* was released in theaters; its plot followed a reporter's discovery of a safety hazard cover-up at a nuclear power plant. Starring Jane Fonda and Jack Lemmon, the film garnered four Oscar nominations and captivated the attention of a nation terrified by the stories of the Three Mile Island meltdown.[21]

In this climate, test site downwinders stepped onto the national stage to share their stories in the joint congressional hearings on the "health effects of low-level radiation." The hearings began in Salt Lake City's Salt Palace on 19 April 1979. Governor Matheson testified first, presenting "many pounds of documents which purport[ed] to show the federal government pushing aside the fears [of health risks] to continue nuclear testing programs." The *Salt Lake Tribune* quoted Matheson: "On the whole . . . I am convinced that the sacrificial lamb in the years of government inquiries has been the truth." The paper reported, "As the governor pieced the story together, he believed the AEC people were decent, sincere people at the outset." When his inquiry brought the AEC documents to light, Matheson concluded the AEC had deliberately "covered up public access to data on fallout and potential danger, and simultaneously was adding to its knowledge about apparent dangers of exposure."[22]

Matheson's documents, along with testimony from former AEC scientist Harold Knapp and Dr. Joseph Lyon, helped to establish the now-defunct AEC as a secretive, manipulative, and arrogant character in the hearings, despite the best efforts of the federal officials who testified on the AEC's behalf. At one point cochair senator Edward "Ted" Kennedy (D-MA) described the AEC as having a "flippant and sarcastic posture" toward safety and downwind citizens' concerns.[23] The AEC as such had ceased to exist in 1974, when the agency's responsibilities and staff were divided between two new agencies—the Energy Research and Development Administration and the Nuclear Regulatory Commission—that were folded into the Department of Energy in 1977. Mentions of the AEC during the hearings and subsequent lawsuits seemed to refer to a defunct character from a previous era, shielding the DOE from negative association with the commission's mistakes.

Even without the stories and evidence presented by and on behalf of the downwinders, the AEC's standing was likely doomed to condemnation from the start in the hearings. The alarm surrounding the Three Mile Island accident had the nation on edge about nuclear danger, and Three Mile updates—such as the *Salt Lake Tribune*'s "Stricken Reactor Cools below Boiling Point, Months from Cold Shutdown Condition"—were front-page news during the hearings. Committee members and readers of the *Salt Lake Tribune* learned, for example, that a single nuclear test conducted in Nevada in 1953 had emitted a radiation dose "40 to 500 times higher than that which triggered the evacuation . . . after the Three Mile Island nuclear reactor accident in Pennsylvania." These headlines amplified the resonance of downwinder testimony, which followed the testimony of scientists and medical researchers.[24]

In considering the 1979 hearings and the downwinder and uranium stories at large, it is tempting to dismiss these citizen storytellers as offering only anecdotal evidence, since they had no formal training as scientists or doctors. They had a different sort of expertise, however, one just as crucial in reconstructing the effects of radiological pollution—local knowledge. Formally educated physicists and medical doctors could offer complex explanations of the processes inherent in nuclear reactions and disease patterns, but these same experts had little to no concept of the structure of local watersheds, foodsheds, and wind patterns, information that many who lived downwind knew intimately. Expert epidemiologists could construct models to chart regional disease patterns, but most would do so based solely on diagnoses listed on death certificates. That paperwork indicated nothing about lifestyle factors affecting exposure, factors local residents could recite from memory.

Consider the theoretical example of a middle-aged farmer named

Matthew Smith, whose death certificate listed heart failure as his cause of death. A professional epidemiologist surveying radiation-related disease in the region would probably pass over Matthew Smith. Smith's wife, however, could tally the half-dozen cancer battles her husband had survived before his heart gave out. Her husband's farmhands could speak authoritatively about the number of hours he spent outdoors working in his fields, the quantity of local milk and homegrown food he ate, the location where the milk-producing cattle grazed, the origin of the water that irrigated the alfalfa that fed the cattle, the frequency with which dust drifted over that alfalfa after nuclear detonations to the west, and the way the wind deposited that dust over the grazing areas. The Smith family could also list other people in the same area, age group, and field of work who had suffered similar health problems. In short, as many sociologists and folklorists have observed, people are experts on their own lives. When the lives of ordinary people intersect with historical events of national and global significance, those ordinary people often have significant expert testimony to offer on those events if they are given the opportunity to speak.

When they stepped up to the microphone at the 1979 Salt Lake City hearings, residents of the downwind region shared stories and evidence they had been piecing together for decades. Sheep owners Kern Bulloch, Jack Pace, and Annie Corry recounted the government's "humiliating" and disingenuous handling of the 1953 sheep die-off. Bulloch offered the story of how AEC officials told him and his brother they were in a "hot spot" and, months later, that their sheep were "hotter than a two-dollar pistol." Five downwinders told their stories of illness and loss, and other downwinders would echo those stories in subsequent days of testimony as the hearings shifted to Las Vegas and St. George. Their stories shared a set of fundamental themes: a past rich with outdoor living, hard work, and local food; a lack of warning from the government about nuclear danger; descriptions of disease clusters; and intense feelings of betrayal coupled with persistent expressions of patriotism. These themes helped shape the narrative structure of the ensuing press coverage, which in turn helped shape the formation of new downwinder narratives in subsequent months and years.

Many of those witnesses who testified before the hearings were part of a "Committee of Survivors" that Irma Thomas had organized at the urging of Stewart Udall in the fall of 1978. Thomas, of course, had been doggedly collecting information and networking with survivors for nearly a decade and long before journalists and politicians began their scramble to investigate. Not coincidentally, all the officers of the committee—"Irma Thomas, St. George; Loa Johnson, Cedar City; Rose Mackelprang, Fredonia, Arizona, and Martha Bordoli Laird, Carson City, Nevada"—were mothers. So too, were most of the uranium widows organizing on the Navajo reservation.[25]

These women were and continue to be part of a global and historical phenomenon in which mothers are a driving force behind movements for change. Many women who have never taken an interest in politics or social causes find the experience of motherhood fundamentally shifts their perspective. As writer Violeta Garcia-Mendoza related, "The first year of motherhood [was] enough to startle me to action. Suddenly, I no longer [felt] paralysis, or inadequacy in my political roles as mother, as woman, as person. With an expanded worldview, I no longer [felt] unworthy of raising my voice on behalf of the people I love—everywhere. Political action and thought became certain consequences of the maternal." Robert D. Bullard, whose research is often cited as the foundation of the environmental justice movement, observed in 2012: "It's no accident that 75 percent of environmental justice leaders are women. . . . A lot of activism emerges from concerns about family, home, and community, and women are more involved there."[26]

In the latter half of the twentieth century, mothers fueled major revelations about the connections between environmental pollution and disease. As the primary caregivers in their communities, mothers bear painful witness to the suffering of children and other family members. They share their stories with other women in their social networks, often piecing together patterns of illness in their neighborhoods years before men or medical professionals take note. Bullard observes that many activists in the environmental justice movement "are retired schoolteachers or women who were active in the church or the PTA [Parent Teacher Association] but weren't active on environmental issues until they were confronted by injustices where they live. After they get involved, they see that this issue is much larger than the local landfill or incinerator." Fueled by love and a sense of responsibility to future generations, their activism is fierce and tenacious and a potent force for change. An antinuclear activist, Representative Bella Abzug (D-NY) once observed, "All we needed was to tell the mothers of the world that their children were drinking contaminated milk from bottles or breasts. There is nothing more ferocious than a woman protecting her young."[27]

By compiling lists of the affected and deceased, by considering the avenues via which contaminants might have reached their loved ones, and by mapping out the literal lay of the land, these maternal activists compiled the evidence needed to convince those in power that a closer look was merited. The Committee of Survivors' leadership demonstrated, as the efforts of Women Strike for Peace had in the previous decade, that mothers could exercise more than a reactive protective instinct. They were also skilled community organizers, and as storytellers they could be highly effective at convincing others of the urgency of the crisis at hand.

While their testimony could bring other mothers on board and compel journalists and politicians, mothers and female activists found many medical and scientific authorities dismissed them on the basis of their femininity. Prior to Bethany Peterson's diagnosis, doctors frequently told Claudia Peterson she was "overreacting" to her daughter's severe leukemia symptoms, suggesting the child was "constipated," "a little anemic," or experiencing "growing pains."[28]

For centuries, mysterious ailments experienced or observed by women have been medically attributed to female emotional weakness and dubbed hysteria, or housewife syndrome. Doctors confronted with radiation-related ailments in downwind communities applied the hysteria diagnosis frequently. In 1954, Iron County agricultural agent Stephen Brower accompanied a Public Health Service doctor on a trip to investigate the health problems of a southern Utah family in Hamlin Valley near the Utah-Nevada border. Brower recalled, "The wife, a young woman in her late twenties, had lost all of her hair, had experienced lesions on her ears and nose that were extremely slow in healing, and she lost her fingernails"—all likely signs of serious radiation exposure. In addition to her physical symptoms, the woman "developed some serious emotional problems . . . and was unable to take care of her children, which resulted in a divorce." Brower later recounted that "the official explanation that was reported to me afterwards was that she had undergone [a] hysterectomy and this explained the symptoms."[29]

When Darlene Phillips of Bountiful, Utah, began to experience severe health problems in the late 1950s and early 1960s, eventually "collaps[ing] with hepatitis in '62," doctors insinuated "it was just more of Darlene being 'a bit nutsy.'" She recalled that one doctor who treated her for pneumonia "put on my chart 'housewife syndrome.' That diagnosis followed me from doctor to doctor, and everybody saw it so they'd say 'This lady's just making up symptoms.'"[30]

In considering the hysteria diagnosis, Salt Lake downwinder Mary Dickson observed matter-of-factly, "I mean, [women are] losing their babies. Of course, they're hysterical. . . . But you're not allowed to be. Especially if you're a woman, you're not allowed to be, because already they're not going to pay attention to you—you're a woman. If you get hysterical, no one listens." Dickson's perspective is informed by decades of speaking out and testifying on behalf of downwinders. "I learned really, really early on that if you want to be taken seriously, you have to remain very calm and almost unimpassioned when you make your case. Especially if you're a woman. The minute your voice goes up, the minute you start sounding hysterical, or you look like you're gonna cry, or you just cry, they stop listening." This experience conditioned many downwinder women to adjust their narratives as they related them to new audiences. They began striving to suppress emotion and incorporate as many verifiable facts as possible.

As the April 1979 hearings progressed, Stewart Udall and Tucson lawyer Dale Haralson busily prepared class action lawsuits on behalf of test site downwinders and uranium-affected plaintiffs. Lawyer Dan Bushnell was also hard at work drafting an appeal for the twenty-three-year-old decision in *Bulloch v. United States*, based on new evidence uncovered by the Matheson commission. *Irene Allen et al. v. United States* was the first suit filed, on 30 August 1979. Chosen to represent the widest possible range of cancers, twenty-four downwinder plaintiffs from Utah, Arizona, and Nevada served as test cases, and close to twelve hundred plaintiffs were waiting in the wings. Their unique challenge was proving the government's liability. "Such responsibility," Frank Fischer explains, "lies with the affected parties rather than the potential polluters. A legacy of industrial society's faith in progress, the principle is institutionalized in the legal system's assumption that industrial production will be benign unless demonstrated otherwise." In other words, it was not up to the potential polluters to prove their activities were safe; instead, it was up to the potential victims of that pollution to prove how they had been harmed.[31]

The *Irene Allen* complaint alleged "the defendant knew, or should have known, of the existence of radiological hazards arising from atomic/nuclear experimentation." Furthermore, the attorneys argued, "the defendant owed a continuing duty to use reasonable care to identify and warn civilians . . . thereby allowing these civilians to fairly evaluate the risk to them of atomic experimentation and take adequate precautions to protect themselves from, or avoid the danger of, exposure to radioactive fallout." Judge Bruce Jenkins agreed.

He declared in his final opinion: "At the core of this case is a fundamental principle, a time honored rule of law, an ethical rule, a moral
tenet: 'The law imposes a duty on everyone to avoid acts in their
nature dangerous to the lives of others.'" Jenkins decided *Irene Allen*
in favor of the plaintiffs in 1984, awarding them damages of nearly
$2.7 million. Their victory was short lived; the government prevailed
in an appeal, overturning the *Irene Allen* decision.[32]

Dan Bushnell succeeded in reopening *Bulloch vs. United States* in
February 1981. In a historic decision, Judge Sherman Christensen,
who had presided over the 1955 lawsuit, vacated his twenty-six-year-old
verdict and found in favor of the ranchers, concluding the government had perpetrated "a species of fraud upon the court." The Tenth
Circuit Court in Denver subsequently overturned Christensen's new
verdict, dismissing his accusations of government fraud and citing
a lack of evidence that radiation exposure from the NTS caused the
sheep deaths.[33]

It was part of a discouraging pattern emerging for downwinder and
uranium-affected plaintiffs, whose legal victories were consistently
overturned on appeal in higher courts, based on a technicality in federal tort law. Prior to 1946, the federal government had been protected
from civil suits by a doctrine rooted in antiquated law known as the
doctrine of *sovereign immunity*. This doctrine held that the powers
that be—whether pope, king, queen, or president—were incapable
of committing illegal actions and were thus immune from lawsuit.
In 1946, the U.S. government abolished this doctrine and opened
itself up to lawsuits with the passage of the Federal Tort Claims Act.

Liability arises from the failure to exercise reasonable care. The
Federal Tort Claims Act established that the federal government
would be liable for failing to exercise reasonable care in a given situation if a private party would be liable in that same situation under
the law of the state in which the situation occurred. The government
could now be sued, but formidable exemptions had been put in
place to protect it. Chief among them was the discretionary function

exemption, which "provides·that there is no liability for the exercise or performance or the failure to exercise or perform a discretionary function or duty." In 1954 the case *Berkovitz v. United States* defined *discretionary functions* as actions taken by government officials who had the choice and the opportunity to "exercise policy judgment," a broad definition that could be applied to many different actions. While *Berkovitz* provided judges with some guidance, the scope and meaning of the discretionary function exemption remained confusing and controversial, giving government lawyers a huge advantage in litigation.[34]

Uranium activists were bumping up against similar liability riddles in their quest for legal reparations. Timothy Benally recalled, "First they [tried to] sue the mine operator, and they told them that they could not sue them because they were not the ones that used the uranium. So they went after the [Department of Energy]. And they too were not suable. Unless they said, 'Okay, [we] can be sued.'" After almost five years of preparation, *Begay v. United States*, Udall's case on behalf of the uranium families, went to trial in August 1983. At the core of the plaintiffs' argument was the contention that the 1952 Public Health Service study conducted on miners without their knowledge or consent constituted a violation of the Nuremburg Code. As the case was being prepared and tried, epidemiological evidence continued to emerge in support of the plaintiffs' case. For instance, in March 1982, the *Navajo-Hopi Observer* reported that researchers had found higher rates of birth defects in mining areas. The following month Leon S. Gottleib and LaVerne A. Husen's study of "Lung Cancer among Navajo Uranium Miners" appeared in the medical journal *Chest*.[35]

While the judge in the *Begay* case concluded the uranium workers' plight "cries for redress," the case was ultimately resolved in favor of the government, which claimed to have withheld warnings about health risks because the uranium was needed for national security. Additionally, the judge "dismissed the Nuremburg Code as having no applicability in U.S. courts" and concluded "that those who carried out the uranium miner study were not 'experimenting on human beings' but rather were 'gathering data' so they could establish safety standards in the uranium mines." In reviewing the finding in 1991, Dr. George Annas offered the assessment that the judge in *Begay v. United States* "adopted one of the Nazi defenses as legitimate: in times of national emergency, research rules must take a back seat to national security." When the state of emergency persists indefinitely, as the Cold War did, this line of reasoning can lead to an indefinite suspension of civil liberties and citizen and environmental protections.[36]

Judges of the Ninth Circuit United States Court of Appeals who heard the *Begay* case explained their decision to uphold the district court's ruling in favor of the government by citing the *Irene Allen* case's explanation of the discretionary function exemption. They wrote, "Suppose a high-level decision maker says, 'International pressures make open-air atomic testing highly necessary. Time is of the essence. We know as a result of such testing some people are going to get hurt. We can't tell them they are going to get hurt. We can't even warn them what to do to minimize or prevent the hurt. In order to preserve our way of life some people unknown to them and unknown to us are going to give all for the good of all.'" This quote from the *Irene Allen* case, cited in a footnote on the Ninth Circuit's overview of findings in the *Begay* appeal, is followed by the words: "Policy decision? Yes. Tort Claims Act exception."[37]

Like the initial legal verdicts in the *Irene Allen* case and the reopened *Bulloch* case, the inception of new epidemiological studies seemed promising for downwinders and uranium-affected people. These studies seemed likely to uncover crucial evidence that activists could use in their efforts to gain recognition of the problem of radiation-related illness, as well as appropriate health care and compensation. Invariably, the studies ran into funding trouble. Scientists found their results dismissed on technicalities, and some felt their careers were threatened when their research indicated a connection between disease rates and nuclear activity.

In a 1997 congressional hearing on "health risks associated with atomic tests in the 1950s" (which proved strongly reminiscent of the 1979 hearings on the "health effects of low-level radiation"), Dr. Joseph Lyon testified about the leukemia study he began in 1977. He stated, "We found excess leukemias that extended into northern Utah where you've got . . . about 80 percent of our population." Lyon testified that subsequent efforts to follow up on his study, widely regarded as the authoritative examination of childhood leukemia related to testing, "thrust [him] into a political situation" so intense that even President Jimmy Carter's personal guarantees failed to procure continued funding for the study. Eventually funding came from the National Cancer Institute, the DOE, and the Department of Defense (DOD). DOE and DOD reviews of Lyon's data omitted his findings of leukemia increases in northern Utah, "confin[ing the data] to southern Utah." Even with the circumscribed findings, his study was still plagued by controversy. Lyon explained: "When the study was slated for publication in the *Journal of the American Medical Association*, someone at NCI or DOD or DOE leaked the results . . . with the statement, 'the findings were inconclusive and did not support the earlier associations.' We were embargoed by the *Journal* and could not respond to those criticisms." He told the 1997 hearing how the DOE and DOD even went so far as to change the numbers in the study: "The advisory . . . committee with the Department of Energy and Department of Defense representatives, very busily revised all the dose estimates downward for northern Utah after they saw the study findings." Lyon explained, "Now this is generally something that graduate school students are failed out of graduate school for, but the federal government was able to get away with it."[38]

Lyon's experience was not unusual. Since the inception of domestic nuclear testing, numerous tactics have been utilized to alter or obscure potentially alarming scientific data about citizens' exposure to radiation. For starters, official estimates of the number of down-wind citizens potentially endangered by NTS activities have changed regularly since the site was first proposed. Frederick Schoemehl observes that during the site selection process, "Los Alamos scientists invented the term 'sector of safety' to describe the potential fallout area immediately downwind of the proposed Nevada site. But then [those scientists] and the AEC constantly changed the sector's size—a manipulation that had the effect of making a relatively sparsely populated area seem even emptier." As operations at the NTS commenced, Schoemehl contends that "national security concerns and science became entangled," causing "certain research agendas" to be privileged over others. "This selective approach, defended in its day by all but a few prescient scientists, led to grave consequences. . . . By insisting that strontium-90 was the only fallout constituent to pose a danger to humans . . . the AEC discounted other byproducts" equally threatening to human health. Central among them, Schoemehl argues, was iodine-131, which rapidly infiltrated the agricultural landscapes and human food chain of the Intermountain West.[39]

The selectivity of scientific measurements of both populations and pollution in the early test site's years bears out sociologist William Bruce Cameron's observation that "not everything that can be counted, counts. Not everything that counts can be counted." Lay citizens often perceive scientists as objective researchers bound by some sort of ultimate truth, yet the data scientists choose to gather and the conclusions they choose to draw are subject to numerous mitigating factors, and chief among them is the entity funding the research. It is not safe to assume that "everything that counts" will in fact be studied, legitimized as statistically significant, or made public. Phil Brown and Edwin Mikkelson explain that "traditional epidemiologists prefer false negatives (type II error) to false positives (type I error); that is, they would prefer falsely to deny an association between variables [e.g., a connection between fallout and disease] when there is one than to claim an association when there is none." Scientist

Beverly Paigen, who worked with citizens of Love Canal, suggests this preference between type I and type II errors depends "on what one perceives to be the consequences of making the error. To conclude that something is real when it is not means that a scientist has followed a false lead . . . this may be embarrassing and harmful to the scientist's reputation." The alternative course, in which a scientist "ignore[s] the existence of something real means that a scientist fails to make a discovery. This may be disappointing, but it does not harm a scientist's reputation, so the scientist is more willing to make type II errors." Paigen reflected that prior to her work at Love Canal, "I also needed a 95 percent certainty before I was convinced of a result. But seeing this rigorously applied in a situation where the consequences of an error meant . . . miscarriages, stillbirths, and children with medical problems, I realized I was making a value judgment."[40]

Frequently those who choose to err on the side of type I errors in the hopes of protecting human health are ostracized by others in the scientific community and are discredited for making what are perceived to be reckless claims. In his 1981 deposition on behalf of plaintiff Dave Timothy, Dr. Robert Pendleton claimed he had been bullied and discredited by the AEC, the Nuclear Regulatory Commission, the Energy Research and Development Administration, and even the Public Health Service in response to his warnings about iodine-131 contamination of Utah milk. "I was quite vocal about the potential hazards," he stated, suggesting his research "stirred up a hornet's nest." He described losing funding, witnessing the disappearance of research data, and being excluded from other studies related to downwind pollution that were frequently dependent on his early research and dosimetry estimates. He added, "I have been told . . . that I'm senile, and [that in the] past that I didn't know what I was doing . . . there is a generalized feeling that I talked out of turn and I ought not to have done that."[41]

Testifying in 1997 Dr. Lyon commented on the government's decision to hide the Weiss leukemia study in the 1960s: "It became instructive, because that's generally how the federal government responds to issues of citizens' concern, i.e. the negative [correlation between radiation and disease] was strongly emphasized, the positive [correlation between radiation and disease] was suppressed or covered up, and those who found the positive were frequently indicted." By Devra Davis's account, the experiences Pendleton and Lyon describe are not uncommon. She cites the stories of numerous pioneering scientists—Mary Amdur, John Goldsmith, and Lester Breslow—whose indictment of industrial pollution as a threat to public health negatively impacted their funding, reputation, and ability to publish. They even were subject to intimidation and threats. In Davis's estimation, the objectivity of scientific institutions is a myth. "Not everybody plays fair," she observes.[42]

However gratifying it may have been to see journalists, judges, and elected officials finally discussing their plight, downwinder and uranium activists knew better than to hand over their struggle for recognition, compensation, and assistance. The ultimate failure of their lawsuits and the crumbling of numerous scientific studies only confirmed this hunch. While they realized public recognition of their crisis and a legal and political remedy were essential, activists such as Irma Thomas, Manny Pino, and Phil Harrison faced a continual struggle on the ground level in their communities, where the "complete denial" Pino had observed still tended to reign supreme. Illnesses were still appearing, people were still dying, nuclear tests were still being conducted underground at the NTS, and uranium was still being extracted and refined across the American West.

As the 1980s unfolded, the uranium industry was on the wane. The Three Mile Island accident, the stories of the downwinders, the work of increasingly well-organized antinuclear activist networks, such as the Clamshell Alliance, and the 1986 meltdown at the Chernobyl nuclear power plant in Ukraine all gave citizens grave concerns about nuclear power. While the decline of the uranium industry boded well for the health of uranium workers, toxic radiological pollution continued to enter the environment from abandoned and unreclaimed uranium industry sites across the Four Corners region and elsewhere in the West. In this region economic hard times, which many had staved off with uranium industry paychecks, returned, and health problems continued to emerge.

Despite the passage of the 1977 Surface Mining Control and Reclamation Act and the establishment of the Uranium Mill Tailings Remedial Action Project, which in 1978 authorized funds and oversight for the cleanup of abandoned uranium mines and tailings piles, the cleanup process proceeded agonizingly slowly, particularly on Indian reservations. Robert Bullard notes that nationwide, "it takes longer to get contaminated sites in communities of color added to the official Superfund list of places to be cleaned up." He explains, "When remediation does come, it's done on the cheap: instead of digging up the contamination and putting clean soil down, they place a fence around the site. That's nothing but a band-aid." As the 1980s and 1990s progressed, more and more residents of uranium-affected regions realized that they were surrounded by toxic waste, and the government that had sanctioned the creation of that waste and the companies that had profited from it were in no hurry to clean up their messes.[43]

Many rural downwind and uranium-affected communities still lacked access to effective cancer detection and treatment facilities. Those families who suffered the loss of a breadwinner to cancer were in immediate and dire economic straits. To address these urgent issues, activists began to form citizen action groups dedicated to providing information about common radiation-related ailments, to rallying support, and to opposing continued weapons and uranium development. Many of these groups emerged from the citizen networks that had been established informally years before, such as Irma Thomas's network of mothers who maintained and shared lists of new illnesses and deaths in their communities. Utah downwinder Janet Gordon and others formed the group Citizens Call in 1978.

Their goals, along with spreading information about the health risks of nuclear exposure, were gathering "[d]ocumentation of personal experiences with radiation fallout in Utah" and "attempting to stop underground nuclear testing at the Nevada Test Site." In 1980, Preston Jay Truman founded the organization Downwinders and pursued similar goals. That same year Bennie Levy helped form the Nevada Test Site Workers Victims Association, and Phil Harrison, Timothy Benally, Harry Tome, and a group of uranium industry survivors formed the Uranium Radiation Victims Committee.[44]

Citizen advocacy groups and other alliances formed in the late 1970s and 1980s persisted for decades, giving rise to or morphing into other, similar advocacy groups: Citizen Alert Nevada, the Healthy Environment Alliance of Utah (HEAL Utah), Downwinders United, Downwinders.org, the Shundahai Network, the Office of Navajo Uranium Workers, and the Southwest Research and Information Center. As radiation-affected people lost legal battles on appeal, one by one, in the mid-1980s and the interest of politicians and journalists came and went, these advocacy groups kept fighting for the funding, completion, and publication of epidemiological studies; to keep the cause of the downwinders and uranium-affected communities in the public eye; and to help those who fell ill obtain basic medical care and assistance.

Eventually, their ceaseless work and testifying bore fruit. In 1990 the Radiation Exposure Compensation Act (RECA), first proposed in 1981 to provide "compassionate compensation" for cancers and deaths related to radiation exposure, was finally made law. While it has been amended several times to expand the group of ailments covered and the citizens who are eligible for compensation, many in downwind and uranium-affected areas are still ineligible. There are five categories into which applicants to the current Radiation Exposure Compensation Program (RECP) must fall. Three emerge from the uranium industry, which had locations in Arizona, Colorado, Idaho, New Mexico, North Dakota, Oregon, South Dakota, Texas, Utah,

Washington, and Wyoming. Miners who suffered from silicosis, pneumoconiosis, cor pulmonale related to fibrosis of the lung, pulmonary fibrosis, fibrosis of the lung, or lung cancer who can provide proof of one year of employment in uranium mines between 1 January 1942 and 31 December 1971 may qualify for a $100,000 payment, which may be disbursed to their survivors if the worker is deceased. Uranium millers or ore transporters with one year of work history who became ill with lung cancer, pulmonary fibrosis, fibrosis of the lung, renal cancer, silicosis, cor pulmonale related to fibrosis of the lung, pneumoconiosis, or chronic renal disease may also qualify for the $100,000 dollar payment. As of 2014, RECP applicants are permitted to combine periods of work in uranium milling, ore transporting, and mining to make up the required one year of work history.[45]

"Onsite Participants" are defined as individuals who worked at the "Pacific Test Sites, Nevada Test Site, South Atlantic Test Site, Trinity Test Site, any designated location within a naval shipyard, air force base, or other official government installation where ships, aircraft, or other equipment used in an atmospheric nuclear detonation were decontaminated; or any designated location used for the purpose of monitoring fallout from an atmospheric nuclear test conducted at the Nevada Test Site." They (or their survivors) may qualify for a $75,000 payment if they have been diagnosed with one of the following diseases: leukemia (but not chronic lymphocytic leukemia); multiple myeloma; primary cancer of the pharynx, thyroid, pancreas, female or male breast, esophagus, bile ducts, salivary gland, small intestine, brain, stomach, urinary bladder, colon, lung, ovary, or gall bladder; lymphoma other than Hodgkin's disease; or primary cancer of the liver, except if cirrhosis or hepatitis B is indicated.[46]

"Downwinders" are defined as individuals who can demonstrate having resided in certain counties in Utah, Nevada, and Arizona between January 21, 1951, and October 31, 1958, or between June 30, 1962, and July 31, 1962. Those counties include Utah's Beaver, Garfield, Iron, Kane, Millard, Piute, San Juan, Sevier, Washington, and

Wayne; Nevada's Eureka, Lander, Lincoln, Nye, White Pine, and that portion of Clark County that consists of townships 13 through 16 at ranges 63 through 71; and Arizona's Apache, Coconino, Gila, Navajo, Yavapai, and that part of Arizona that is north of the Grand Canyon. Compensable diseases for downwinders include leukemia (but not chronic lymphocytic leukemia), multiple myeloma, lymphoma (other than Hodgkin's disease), and primary cancer of the pharynx, small intestine, thyroid, pancreas, male or female breast, esophagus, bile ducts, salivary gland, brain, stomach, urinary bladder, colon, gall bladder, lung, ovary, or liver (except if cirrhosis or hepatitis B is indicated). Individuals who meet all of these criteria, or their survivors, may qualify for a payment of $50,000.[47]

It is important to consider the RECP guidelines in all their details because one can read much between the lines. The government's specifications reveal some picture, however limited, of the areas it essentially admits to having exposed while offering those who do qualify the small comfort of what amounts to an admission of responsibility. For those applicants who fail to qualify but can see little difference between their suffering and that of a neighbor who did qualify, the RECP guidelines are a slap in the face. Not only do they not receive any compensation, but also the government offers no sense of apology and takes no responsibility for their exposure.

The very notion that certain cancers are attributable to radiation while others are not is absurd to most downwinders and to scientists such as Dr. John Gofman, formerly a top medical researcher for the AEC. In 1992 Gofman lambasted the compensation guidelines as "absolute lies. . . . Basically, I wouldn't believe anything written by the Department of Defense or the Department of Energy. And the North American Radiological Association is worthless." Gofman pointed out, "The government holds that only [certain] cancers are radiation-induced and therefore there's no compensation for other cancers." The veteran researcher dismissed the list of radiation-induced cancers as "nonsense. I regard all cancers to be equally inducible by radiation, so I would consider that bill as erroneous." Claudia Peterson articulated the suspicions of many downwinders when she suggested "they took skin cancers off" the list of compensable diseases "because there were too many test site workers" who could have filed claims for skin cancer. Equally absurd is the notion that radiation respected such arbitrary boundaries as county and state lines. While government-affiliated epidemiological studies, most notably the National Cancer Institute's 1997 thyroid cancer study, have pinpointed hot spots and testing-related cancer clusters in Idaho, Montana, and the Midwest, the downwinder portion of RECP remains limited to a tiny area in the Four Corners region.[48]

While downwinders acknowledge compensation is appropriate and has certainly aided many who received it, many activists find RECA so inadequate as to be insulting, partially because the existence of such a program insinuates historical wrongs have been righted. Downwinder activist Jay Preston Truman refers to the compensation as "blood money." He refuses to accept the payment (for which he qualifies) and has demanded coverage for all downwinders under the RECP, writing, "JUSTICE, not JUST US."[49]

Expanding the RECP has become the central aim of many in the downwinder activist community. The 1997 NCI report on thyroid cancer galvanized many outside of the covered RECP region when it revealed that the counties with the heaviest I-131 exposure were actually a good distance from the Nevada Test Site. Meagher County, Montana, ranked highest, followed by Custer County and Gem County, both in Idaho. Downwinder stories proliferate in these regions, where activists such as Tona Henderson have been piecing together their own families' exposure stories and collecting those of their neighbors. Just as Irma Thomas did, Tona documents every new disease diagnosis and death in her community that are suspected of being linked to radiation exposure. Her file runs to nearly a thousand names, and her voice mail and e-mail fill weekly with new stories. Years of organizing work and lobbying by the mother of five and her fellow reluctant activists across the western states have pushed the issue back into the political arena yet again. Senator Mike Crapo (R-ID), Senator Tom Udall (D-NM), and others have repeatedly introduced legislation seeking to increase the compensation amounts and expand the list of covered areas. They hope to add Colorado, Idaho, Montana, and New Mexico to the roster of RECP states, as well as the those counties in Arizona, Nevada, and Utah that are currently excluded.

Activists in uranium-affected areas are similarly focused on reforming RECP. Phil Harrison and Timothy Benally both note that compiling the necessary RECP paperwork to apply in the first place discourages many, particularly on reservations, from ever completing the process. Many of the uranium-affected Navajo families Susan Dawson interviewed "found the social service and legal systems too complicated with bureaucratic entanglements and so did not access them at all, or gave up on them during the process." Upon being asked why she did not apply for compensation, one widow told Dawson that "she felt intimidated by the process because of being told she had to write letters. She had no stationary or stamps and could not write in English, and so [she] decided against it."50

The application process is one of several major roadblocks. In order to substantiate early RECP claims for uranium workers, advocates realized they needed trained medical personnel to administer medical evaluations. In the early 1990s Timothy Benally became a caseworker for the Office of Navajo Uranium Workers. He encountered "a problem right from the start. We didn't have any qualified doctors to do the [required] pulmonary function testing." His office contacted the Public Health Service, which "sent us a doctor [Dr. Louise Abel] who was not a real expert in the area, but she was good in diagnosing a lot of the work we had to do." Abel could only assist RECP applicants through portions of the process. Benally and his fellow advocates "found that there were a lot of other areas that needed attention from somebody," and their struggle to gain assistance in negotiating the application process continued.

Phil Harrison recalled how in the first years after the RECP was signed into law, "people were complaining about they're not getting paid. The criteria was so stringent . . . nothing was happening. So in 1995 we started a process of amending the laws." In 2000 activists and political advocates succeeded in pushing through an amended version of the Radiation Exposure Compensation Act, which expanded compensation to new groups of radiation-affected claimants. Harrison said, "That's when the Navajo downwinders [Navajos and other indigenous residents of northern Arizona who were directly in the path of test site fallout] were added, the truck drivers, the mill workers, the surface miners." Despite the program's expansion and refinement, before long Harrison began to hear "another public outcry about people not getting claims, not getting paid."

Harrison and other activists traveled to Washington DC to learn why so many Navajo claimants were not being compensated. "It came to the fact that it was because of [Navajo claimants lacking] records, no original records, marital records, birth certificates. . . . They wanted these originals." While many European American RECP applicants have difficulty finding their fifty-year-old birth certificates, sixty-year-old marriage certificates, or twenty-year-old medical records, it is next to impossible for many indigenous RECP applicants to locate such documents, since many marriages, births, and deaths were not formally documented on reservations decades ago. Harrison believes the insistence on official original records is "uncalled for. We do not have to produce those things. . . . If the chest x-ray is positive, if the doctors diagnosed them with lung disease, that should be it, and work history, minimum work history [for uranium workers and their families]. People should get paid." Harrison continues to advocate for additional reformation of the RECP. When interviewed in 2008 he was pushing for an increase in compensation award amounts and continuing to speak on behalf of uranium workers whose health problems were excluded from compensation.[51]

Lorraine Jack also described the application process: "People . . . reaching out to receive the money, but they cannot reach it. It is almost like someone has money in their hands and the people are chasing around after them . . . in the process, they are stumbling and hurting themselves while they are chasing after the money." Jack spoke of friends and relatives who "pawn[ed] their things to have gas money to go to the hospital, and they would like to use this [compensation] money to get their pawns out. They hope for this, and they [the government] seem to be giving them a difficult time and it is not good." Even for those who do obtain compensation, Jack observes, the money fails to bring much comfort compared to what they have lost. "We have all been exposed to uranium, and then the gratitude of one hundred thousand dollars is just not enough. The people have been contaminated, it is in their blood, and some children never got to see their fathers come home to them. Money is something that just disappears in your hands, and human beings are priceless. Human beings can greet you all their life. Money does not greet you."[52]

It is tempting to make out RECP administrators as bureaucratic villains, denying downwinders and uranium-affected families their due at every turn. In actuality, many of the Department of Justice lawyers and doctors affiliated with the Radiation Exposure Compensation Education Program (made possible by grant money included in the 2000 amendment of the RECA) are dedicated advocates on behalf of downwinders seeking compensation. Many, such as Dr. Douglas Zang, chose to do the work out of a passion for social justice. As a doctor in the education program, Zang works to inform downwinders, uranium workers, and health care professionals about the health risks associated with uranium exposure, as well as providing screening services for those who may qualify for compensation. He realizes that not everyone he attempts to advocate for differentiates between him and the agencies that have failed to protect them. "I'm in this interesting position," he mused in a 2008 interview.

The people were betrayed by the Public Health Service in the Colorado Plateau study. I work with the Indian Health Service. The compensation program is administered by the Department of Justice. As far as people are concerned, it is all Washington. I've used these terms: Public Health Service, Indian Health Service, Department of Justice, Department of Labor, Health and Human Services. When the interpreter, who does this every day, translates any of those terms, it's "Washington." It's one word for all those things, because in people's minds, it's all one thing. . . . I am Washington, the Public Health Service is Washington, and the Department of Justice is Washington.[53]

Zang's awareness of the history of the community he lives and works in, as well as his sensitivity to patterns of environmental and medical racism, is crucial to bridging this sort of mistrust. Efforts also are under way to train new doctors to bring a similar sensibility to their practices. In January 2004, the Southwest Information and Resource Center partnered with the University of New Mexico's Family and Community Medicine Department to put on a two-day tour of uranium-affected communities in the Eastern Navajo Agency as part of a "training program designed to give new physicians real-world examples of what is meant by the terms 'environmental health' and 'environmental justice.'"[54]

Well-known community activists such as Manny Pino and Claudia Peterson also are often called upon to assist others in navigating the daunting process of RECP applications. For some of these applicants, the act of pursuing compensation represents a major step in acknowledging the government's betrayal of their trust. Pino described coming "full circle" with some of the miners he had attempted to warn decades before: "A lot of those miners that were telling me 'get that shit out of my face' [when presented with information on health effects of uranium], twenty years later, we're helping them file claims under RECP." Pino recounted how many of the men "come up to me with tears in their eyes and they'll put their arms around me and they'll say, 'You know, son, you were right.' But a lot of good that does today, you know? When you have someone who is on their deathbed, and their only wish is that they get this money to help the future of their families."

On the other end of the spectrum are those applicants who pursue compensation but fail to undergo significant politicization or even visible attitude shifts in the process. In a 2005 interview, Claudia Peterson spoke of her exchanges with many RECP applicants in St. George. "They're calling me and they'll say, 'Well, can you help me get my money?' And I'll say, 'Yes,' and I'll say, 'You might want to call up your congressman or senator or make a . . .' 'Oh, no. All I want is my money. Don't put my name . . . don't put me out there or anything.'" Peterson found the reactions confusing for many years. "For a long time I thought it was apathy, I tried to figure it out. Figure out why people were so blind. . . . I think a lot of it is shame. It's pretty embarrassing and pretty awful to believe that your own people, your own government who you're proud of, that you've been patriotic and loyal to, has considered you the lowliest segment of the population." She suggested the experience was a phenomenon "almost like incest. People can't talk about it or are afraid of it."

While Peterson has continued to offer her assistance to her neighbors when she is able, she cannot overlook a painful irony: while many in her community qualify for compensation with her help but refuse to speak out on behalf of the downwinder cause, she and her husband received no compensation for their daughter Bethany's death. Though both parents were children in compensable counties during the atmospheric-testing era, their Bethany was not; thus, her leukemia, under RECP guidelines, cannot be connected to the test site. In 2005 Claudia's nephew, who was serving in the military in Iraq, "fell and broke his ankle and had some stomach problems they thought was from his medication. Did a colonoscopy and he's clear full of cancer. Tumors everywhere. He's thirty years old." A second-generation downwinder like his cousin Bethany, he also fails to qualify for compensation.

New data continues to emerge about the ways those citizens exposed to radiation—and their descendants—are at risk for health problems. In 1997 the National Cancer Institute released what it called "a study of unprecedented magnitude" in the history of down-wind radiation exposure reconstruction, "Estimated Exposures and Thyroid Doses Received by the American People from Iodine-131 in Fallout." "Utilizing the limited data available for each of the 90-plus tests between [nineteen] fifty-two and [nineteen] fifty-eight respon-sible for 99 percent of the I-131 released into the atmosphere," the researchers correlated fallout data "with detailed wind patterns, rain-fall patterns, grazing patterns of cows and goats, transfer patterns to milk, [and] milk distribution and consumption patterns." Then it "analyzed [the results] for all 3,000 plus counties in the 48 contigu-ous states for 13 age groups for multiple milk consumption patterns." The investigation concluded this exposure "may have resulted in as many as 75,000 additional cases of thyroid cancer to the children alive at that time throughout the course of their lives, about a 20 percent increase from the expected number of thyroid cancers."[55]

The 100,000-plus-page NCI study has frequently been pointed to as a landmark acknowledgment of the downwinders' plight, as it is one of the first scientific studies to acknowledge that citizens outside of the immediate downwind region were exposed to fallout. Published online with the goal of increasing accessibility, the report includes maps documenting fallout hot spots through the entire country, as well as a fallout calculator that allows any citizen alive during atmospheric testing to estimate their exposure based on their milk consumption. Testimony from several prominent scien-tists suggests the NCI findings actually understated the true extent of downwind illness and remain simply another flawed product of politically manipulated science.[56]

As noted previously, Dr. Joseph Lyon first encountered troubling data about thyroid disease in downwind areas while attempting to complete his study of fallout-related leukemia in the 1980s. He submitted a request for funding to pursue a thyroid study, but the NCI denied it. He later suggested that in his experience, "if the negative [data] look[ed] too threatening," the institute avoided publicizing that data. In his 1997 testimony Lyon suggested "the current administration" of the NCI lacked "a strong public health interest" and necessary "substantial public involvement" in its epidemiological studies, raising significant doubts as to whether the landmark study even came close to revealing the full picture of radiation-related thyroid disease. Epidemiologist Dr. Jan Beyea echoed Lyon's contention that the NCI's results were understated and expressed concern that scientific scrutiny focused too narrowly on specific ailments such as thyroid cancer while overlooking non-cancerous, radiation-related ailments, like "nodularity . . . [and] autoimmune thyroid disease."[57]

Timothy Conner, associate director of the Energy Research Foundation in Spokane, Washington, condemned the NCI for delaying the publication of the findings. "Its researchers some time ago had compiled and analyzed enough evidence to realize that this was more than just a science project . . . thousands of infants and children throughout America had received thyroid doses putting them at substantially greater risk for thyroid cancer. . . . People had a right to know. And had they known, that knowledge may well have made a difference in their lives." Dr. Robert Pendleton would have agreed with Conner wholeheartedly. He did not live to see the release of the NCI study, however understated its conclusions may have been. In 1981, the year before he died, Pendleton testified: "I think that it is absolutely essential that if a scientist has basic information that is directly applicable to the civil health or the public health and the civil defense of the country that he is absolutely derelict in his duty if he does not bring that material to the attention of the people of this country."[58]

Despite the occasional newspaper headline or congressional hearing, such as those occasioned by the 1997 NCI study, public interest in the plight of radiation-affected citizens never lasts long, and those remedies that are instituted are rarely comprehensive. In the twenty-first century, many areas of the American West heavily affected by radiation still have a dearth of local cancer treatment centers. Applicants to the RECP still face a small mountain of obstacles to obtaining compensation money and frustrating limitations in RECP coverage. Other community members continue to deny the existence of risk from radiological contamination, even as they and their loved ones battle rare forms of cancer and illness frequently linked to radiation exposure. Frank Fischer concisely explains this phenomenon, which is known as industrial fatalism: "Possessing neither the ability to identify those responsible for their anxieties nor the mechanism to redress the sources of the problem, those surrounded by such threats tend to retreat to an old and established defense mechanism: they simply choose not to see or hear."[59]

The supercritical level of awareness of radiation-related contamination reached in the late 1970s was of pivotal importance in advancing the cause of radiation-affected people in the United States. The historical collision of maturing citizen activism with the nascent environmental movement, growing mistrust of the government, and such events as the Three Mile Island and Chernobyl meltdowns created a national environment that was highly receptive to the stories of downwinders and uranium-affected people. Their stories and their organizing efforts provided grist for the emerging antinuclear movement, even as some radiation-affected citizens continued to support nuclear weapons development. Their stories and their activism fostered global publicity, congressional attention, legal actions, new disease research, and eventually some "compassionate" monetary compensation. Ultimately, these remedies provided only partial solutions to the ongoing crisis.

Following the passage of RECA and the Comprehensive Nuclear Test Ban Treaty in the early 1990s, many downwinder and uranium activists began to breathe a little easier. While most believed they still had a long way to go in achieving just compensation and recognition, it seemed their stories might become somewhat less urgent. Most were battling ongoing health problems, attending funerals regularly, and growing weary of being in the public eye. Some chose to stop talking to reporters and fellow activists and instead focus on their health and their families. Few were naive enough, however, to believe the test site and the uranium industry were out of commission permanently, a suspicion confirmed by the nation's response to the 11 September 2001 attacks on the Pentagon and the World Trade Center in New York. Americans reeling at the unfamiliar sight of war waged on American soil compared the experience to Hollywood movies and hearkened back to the Japanese bombing of Pearl Harbor. Overwhelming support was given to a president who saw military action as the only appropriate response and who promised, "Whether we bring our enemies to justice or bring justice to our enemies, justice will be done." President George W. Bush's words at the inception of the war on terrorism were eerily reminiscent of President Harry Truman's words at the close of World War II. Both suggested the United States faced a clear enemy who had initiated the conflict and deserved retribution. Not surprising, the resumption of nuclear weapons technology development was part of the plan for protecting national security following 11 September 2001.[60]

The National Atomic Testing Museum, which opened in February 2005 in Las Vegas, made the connection between the Nevada Test Site and the 11 September attacks explicitly: in its final gallery, next to a fragment of the Berlin Wall symbolizing the end of the Cold War, a chunk of metal from the World Trade Center is prominently displayed next to a photograph of a sign hanging on a gate at the NTS that reads "Always Ready," Meanwhile, the museum provided no historical interpretation for the displays and devoted no exhibit space to the downwinders or uranium-affected people, despite the patient lobbying of activists such as Mary Dickson and Robin Ray.

Many downwinders find the museum painful and insulting in its exclusion of their stories. Michelle Thomas toured it in her wheelchair with a *Salt Lake Tribune* reporter in 2005. She described her initial impressions to me several months later: "As you enter the museum, you are greeted by a life-sized poster, person, 'Miss Atomic City.' She looks a lot like Marilyn Monroe. She wears nothing but stiletto heels and a mushroom cloud." Miss Atomic City, winner of a 1950s Las Vegas beauty contest, was modeling the inspiration for a bathing suit a French designer had christened the "bikini," named after the Pacific Island location of the first postwar nuclear test.

Michelle Thomas continued, "The mushroom cloud looms up from her loins and extends up her torso, where it billows beautifully across her bustline and neck. I thought this was particularly symbolic." Thomas described how she used her cane to "point out to onlookers that the mushroom cloud managed to cover those areas of the female body most threatened by the toxic radioactive fallout—uterus, ovaries, stomach, lungs, esophagus, breasts, thyroid, et cetera. The museum curator was ill-amused." Later Thomas reflected, "I don't know when I've ever been so offended and hurt. We were so minimized, what happened to us."

While it is affiliated with the Smithsonian Institution and the Desert Research Institute, which conducts environmental research in conjunction with the Nevada System of Higher Education, the Atomic Testing Museum is also associated with big names in energy and defense technology. Walking through the Lockheed Martin Entry Gallery, the Bechtel Foundation Stewards of the Land Gallery, and Reynolds Electrical and Engineering Co./Edgerton, Germeshausen & Grier, Inc., Honored Underground Testing Gallery makes it clear that interests with a significant financial stake in nuclear technology have made investments to influence the way nuclear history is represented. The museum's message is clear: the weapons developed at the NTS were responsible for ending the Cold War, and the test site's operators are ready to spring into action again to bring a similar resolution to the twenty-first-century war on terrorism.[61]

The museum's linking of the Cold War and the war on terrorism invites other comparisons. While both conflicts created many casualties, both are heavily conceptual, waged on behalf of a citizenry inspired to fear ideas as much as specific nation-states. Since it is impossible to defeat an idea militarily and the citizenry is told those who would promote that idea (terror or communism) are furtive and lurking everywhere, the conflicts offer those in power indefinite license to restrict civil liberties in the name of national security while funneling massive sums of money into military technologies.

In 2006 pro-testing interests came very close to getting their way. Divine Strake, a 700-ton nonnuclear atmospheric weapons test, was scheduled to take place at the NTS in June. The Department of Defense estimated the ensuing mushroom cloud would rise ten thousand feet into the Nevada sky, and while the blast itself would be nonnuclear, the cloud would carry the pulverized dust—and any lingering radiation—from a site that had hosted numerous nuclear blasts mere decades earlier. "Pentagon spokespeople offered the usual assurances" of safety, writer Chip Ward observed, "even while admitting that they had little in the way of data to back them up. Nothing resembling an environmental impact assessment had been done, but the implication was that the Pentagon's word should be good enough."[62]

Downwinders knew better than to accept the government's guarantees of safety and began their familiar campaign anew, calling elected officials, writing letters to editors, staging protests, and mounting legal challenges to prevent Divine Strake from going forward. Ultimately their efforts succeeded, and after several postponements of the test, the DOD finally canceled Divine Strake in February 2007. While the cancellation was a major victory for downwinders, it was only one test; a far broader plan is in place to resume nuclear weapons development. Dubbed Complex 2030 or Complex Transformation by the DOE, the plan seeks to "ensure that U.S. warheads remain reliable into the future, to create a responsive weapons complex that could rapidly produce additional warheads, to save money, and to allow the United States to reduce the size of its nuclear arsenal. None of these rationales holds up under scrutiny," the Union of Concerned Scientists asserts.[63]

The prospect of new nuclear weapons development, coupled with increasing promotion of nuclear energy as a supposedly clean alternative to dependence on foreign oil, provoked a surge of renewed interest in uranium extraction. "With demand increasing," journalist Judy Pasternak noted in 2006, "the price of uranium has climbed to more than $60 a pound. Six years ago, it was as low as $7." Fully aware of the quantity and value of uranium reserves still buried in the lands of the Southwest, mining companies began promising communities in the region that new technologies such as in situ leach mining could remove uranium without causing environmental degradation. *In situ leach mining* "involves leaving the ore where it is in the ground, and recovering the minerals from it by dissolving them and pumping the pregnant solution to the surface where the minerals can be recovered. Consequently there is little surface disturbance and no tailings or waste rock generated," according to the World Nuclear Association. The uranium industry has thus far failed to demonstrate that the technique can be practiced without depleting and contaminating the water table, but some of those eager to see uranium development resumed claim to be convinced of the method's safety. In 2008 Manny Pino noted, "The city of Grants, [New Mexico,] already passed a resolution saying 'we're pro-uranium mining, we're pro-milling, we need the jobs. This is a new method of mining, it's not like conventional mining, it's not gonna tear up the land, we don't need intense labor force, you know, and we're gonna reap the benefits.'" In situ is eerily similar to underground nuclear testing. Because it promises not to produce visible signs of contamination, it can contaminate the ecosystem silently and invisibly, out of sight, and off the radar of those citizens who might otherwise become alarmed.[64]

Some people who remember the ravages of previous uranium booms have stood firm. In April 2005 Navajo Nation president Joe Shirley signed a ban outlawing all uranium mining and processing on the Navajo reservation, a place that Uranium Resources, Inc., vice president Mark Pelizza called "the Saudi Arabia of uranium." In spite of the ban, mining concerns continued to lobby the Navajos, prompting Shirley to pass Executive Order 02–2005, mandating tribal employees "avoid any 'communications with uranium company representatives.'" This mandate "infuriated mining executives," Judy Pasternak reported. She quotes Strathmore Mineral Corporation executive John DeJoia, who rhetorically asked: "You tell me, what kind of a democracy is that? They've got tremendous resources out there. They're a very poor nation. That could change." As of 2014, the Energy Fuels Company was in the process of obtaining permits for the Roca Honda Project, a major uranium operation sited near Mt. Taylor, just outside of Grants, on land designated as a Traditional Cultural Property, owing to its significance to the Acoma and Laguna Pueblo people, the Navajos, and the Hopis. The project faces fierce resistance.[65]

In every area offered promises of economic rewards for acceptance of nuclear technology, there are plenty who would prefer to take the money and jobs, using the rationale that they have already been heavily exposed to radiation and thus may as well get paid. Robert Bullard points out, "It's difficult for people who are unemployed to turn down an offer of jobs, but local residents don't usually get the jobs that are created" by new industrial sites. He offers a hypothetical scenario: "Say a company wants to build a wastewater plant and incinerator in a community. The company tells residents it will create 500 jobs, but out of those, 450 are construction jobs." Of the remaining fifty long-term positions, Bullard suggests the majority would be "for technicians, engineers, and skilled laborers, few of whom live in the area." Bullard observes pithily, "If the company had said up front 'You'll end up with five cleaning jobs,' the community would likely have decided it wasn't worth the destruction of their environment."[66]

So far the Navajo Nation has stood firm. It is one of several indigenous nations stridently opposing uranium development. The All Indian Pueblo Council, representing all nineteen tribes in New Mexico, passed a 2007 resolution banning any uranium mining that would affect sacred sites, and the Havasupai Tribal Council has banned uranium development on the reservation and continues to stridently oppose it elsewhere, particularly in the Grand Canyon area.

The opposition of tribal governments to uranium development is informed by the stories of their own citizens and the stories of other indigenous peoples from around the world. In 1992 indigenous activists from every continent met in Salzburg, Austria, for the World Uranium Hearings. Navajo, Acoma, Laguna, Havasupai, and Western Shoshone activists who lived only hours apart met—some for the first time—and learned that their experiences were not unique. They discovered that the ravages of the uranium industry had taken a similar toll on indigenous communities in Brazil, Africa, Australia, Polynesia, and elsewhere. Activists such as Alveno Waconda and Gloria Lewis, who had traveled from Laguna Pueblo to the hearings, brought this awareness home to their fellow activists, families, and tribal leaders. What had once seemed a local problem was now understood to be part of a global pattern. In May 2007 Manny Pino and other indigenous anti-uranium activists from around the globe presented the Indigenous World Uranium Summit Declaration at the sixth session of the United Nations Permanent Forum on Indigenous Issues held in New York City. Affirming that "past, present and future generations of Indigenous Peoples have been disproportionately affected by the international nuclear weapons and power industry," the declaration called for a "worldwide ban on all aspects of the nuclear fuel chain." The declaration received "tremendous support" from "the Navajo Nation, various organizations from indigenous nations, and Native representatives from Australia, Brazil, Canada, China, Germany, India, Japan, the United States, and Vanuatu."[67]

Crucial to the passage of the uranium bans and the defeat of the

Divine Strake test have been the testimonies of those uranium activists and downwinders old enough to remember previous eras of uranium extraction and nuclear testing. These activists will be needed to uphold the uranium bans and defeat the DOE's Complex Transformation plan, yet with every passing year, more of them succumb to the very illnesses they are trying to prevent in future generations. In a 2005 interview, Salt Lake activist and thyroid cancer survivor Mary Dickson reflected on the illness of a good friend: "I just can't worry about [cancer] until they diagnose her for sure. . . . I have all these other people dying. This Monday, another woman I knew died of ovarian cancer." Dickson is a tireless organizer and a woman of exceptional energy, but as she recounted the litany of recent diagnoses and losses, the strain was evident in her voice. "They're still dying. This story won't go away. It just won't stop. We're trying to do that memorial [to downwinders] in Memory Grove [in Salt Lake City], and I've got all these people on this group and I just barely have the energy to pull them together anymore." She described one attempt to hold a meeting of the organizers. One of the women canceled "because her son-in-law's dying of pancreatic cancer, the guy from Durango had to cancel his trip because his multiple myeloma is back. . . . I'm really worried about him. And I'm thinking, all these people who are working on this are *too sick* to keep doing it."

"People who live here and go to funerals every week, we can't sit here and talk about it like we're talking about broccoli," Michelle Thomas told journalist Lucinda Dillon in 2001. As noted previously, while a teenager, Thomas was diagnosed with polymyositis, a rare inflammatory disease that weakens muscle tissue and has significantly impaired her ability to use her legs. She has survived ovarian cancer, breast cancer, and, most recently, salivary gland cancer. Phil Harrison's kidneys failed in 1999, necessitating a transplant, and chest x-rays have revealed scarring on his left lung. Ivan Sidney reflects that "every day is a good day" since his bout with cancer, but "I know the cancer is sleeping in me. It could come back anytime.

And I don't know if I want to go through that whole thing again. I'm not the same anymore. Eighteen months of chemo can ruin anybody."

Downwinders and uranium-affected people of subsequent generations will have a much harder time developing a collective consciousness. While some may still suspect nuclear testing or in situ uranium mining is at fault for their illnesses, their stories will no longer be able to draw on the potent imagery of the mushroom cloud or on stories of children playing on tailings piles. Local, family-run agriculture has declined precipitously since the nuclear age began, contributing to the increasing alienation of consumers from the production of their food. "The passive American consumer, sitting down to a meal of pre-prepared or fast food, confronts a platter covered with inert, anonymous substances that have been processed, dyed, breaded, sauced, gravied, ground, pulped, strained, blended, prettified, and sanitized beyond resemblance to any part of any creature that ever lived," Wendell Berry wrote in 1989. "The products of nature and agriculture have been made, to all appearances, the products of industry. Both eater and eaten are thus in exile from biological reality." Unlike previous generations, modern citizens have few opportunities to become aware of how the uranium industry and nuclear sites (and industry in general) contaminates their food, air, water, and soil.[68]

The influx of new individuals settling in the southern Utah region has increased the population significantly in recent decades, destroying the intimacy of small communities, diluting collective memory, and complicating the observation of local disease clusters. Additionally, "you've got a new generation, and even if their grandparents died of [cancer], they don't know about it," Mary Dickson mused in 2005, considering St. George, Utah. "And then you've got all these outsiders coming in who never knew about it. So you've got this city that doesn't even know its own history anymore." In these twenty-first-century circumstances, patterns of health problems, agricultural crises, and environmental changes can easily become submerged amid larger populations, increasingly remote food production systems, and alienated relationships with the natural world.

While the radiation-affected people of the twenty-first century may not know their neighbors or where their food comes from or that an in situ uranium mine exists down the road, they have access to a tool Irma Thomas would have loved to have gotten her hands on—the Internet. Digital media allows ordinary citizens to access a wealth of research and data from their living rooms or libraries. A mother whose child has been diagnosed with a rare disease can read reports on the drugs used to treat it, find other parents whose children have gone through the same ordeal, and discover the websites of activists seeking to highlight the connection between radiation exposure and disease. This mother can network digitally with activists around the world, and if she wants it, she has a powerful platform from which to tell her story to millions.

CONCLUSION

With particular details and personal anguish, downwinder and uranium-affected storytellers have testified to the ravages of the nuclear era since its inception in World War II. Their testimony about the costs of their Cold War–era radiation exposure brings home the truth that all wars — no matter how conceptual they may seem — occur in actual places, where actual people live, grow food, and raise children. They remind us that national security is more costly and complicated than we have been led to believe. No matter their politics, many who hear these stories find their understanding of American history unalterably changed.

Radiation-affected people tend to remain a side note in broader discussions of what historians call the Atomic West. On 15 October 2005, at the Western History Association Conference in Scottsdale, Arizona, a panel titled "The Atomic Southwest" made no reference to downwinders or uranium-affected people. Historian Ferenc Szasz made the only mention of the devastating environmental implications of the American West's nuclear complex when he noted that the Trinity site in New Mexico, home to America's first nuclear test, was too heavily contaminated to be made into a national historic site. While radiation-affected people can be seen as non-topical in many discussions of American nuclear development, expounding on the Atomic West without at least alluding to their experiences constitutes a continued marginalization of their stories and communities. In response to my questions posed to the Atomic Southwest panel, Los Alamos historian Roger Meade denied the downwinder stories had any scientific or historic validity.[1]

When they are acknowledged, downwinders and uranium-affected people tend to become the "local color" in the historical narrative of the Cold War American West. They are sprinkled in as archetypal western characters: the eco-conscious Native American, the rugged and earnest rancher, the God-fearing Mormon housewife, the hell-raising woman activist. While some downwinders and uranium-affected people embrace these identities to magnify the resonance of their stories, in other people's retellings radiation-affected people end up as two-dimensional folk characters, pointing in horror at a yellow monster or mushroom cloud on the horizon that none of the rest of us can see. Thus fragmented, they are reduced to little more than their caricatures and their cancers and are rarely given the opportunity to represent holistically, in their own voices, the trajectory and significance of their experiences.

When their stories are considered as documentation of historical events, simultaneously personal and culturally collective, compelling patterns of evidence begin to emerge. The downwinder and uranium stories are the products of a complex, historically constructed system of experiential, local knowledge. In addition to being a record of patterns of disease and environmental change, this local knowledge encompasses vital cultural material, including historical and contemporary citizen perspectives on food production, land use, patriotism, gender, economics, sovereignty, and justice. The stories reveal both the evolution of critical lay perspectives on nuclear testing and uranium extraction and the ways the authors of those perspectives have grappled with a political climate (both national and local) that is preoccupied with security and economic growth and intolerant of dissent. These perspectives evolved in the context of institutionalized bias against the knowledge of rural and indigenous people and against the observations of women, as the narratives of the ranchers, uranium families, and many female downwinders make clear. Their stories also reveal the ways they have come to identify themselves as casualties of a global pattern of dismissal and exploitation, in which military power has been won at the cost of the health and safety of rural and indigenous populations.

Their stories remind us that each citizen is uniquely qualified to comment on the time and place in which they are living. If we divorce the tragedies of these narratives from the cultural process by which narrators pieced together an explanation of those tragedies, we dismiss their expertise, and they fade into vernacular anonymity. Witnesses Ivan Sidney, Claudia Peterson, and others then are regarded as just other alarmists blaming their cancer on radiation, when they are capable of articulating a complex awareness of the avenues by which nuclear pollution infiltrated their local environments and found their families.

What is lost if people do not remember what has happened in their communities? Wendell Berry writes that "good local culture, in one of its most important functions, is a collection of the memories, ways, and skills necessary for the observance, within the bounds of domesticity, of . . . natural law." If people stop telling stories about the events, work, and living that occur in a place, Berry suggests human community will erode just as abused soil does. "If the local culture cannot preserve and improve the local soil, then, as both reason and history inform us, the local community will decay and perish, and the work of soil building will be resumed by nature. A human community, then, if it is to last long, must exert a sort of centripetal force, holding local soil and local memory in place." Berry goes so far as to suggest that "practically speaking, human society has no work more important than this. Once we have acknowledged this principle, we can only be alarmed at the extent to which it has been ignored."[2]

The narratives of downwinders and uranium-affected people reveal the process by which rumors about environmental contamination emerge and spread locally. As residents of agricultural landscapes and communities in a region used for military and industrial purposes, these ordinary citizens bore witness to a dangerous intersection of toxic pollution and the food supply, the effects of which rippled far beyond their neighborhoods and ultimately across the country. Because they noticed, they were able to make the connection later on between health problems and toxic exposure. Because they noticed, they were able to offer a counternarrative to the official fictions circulated by their government, which depicted nuclear weapons development as a controlled experiment crucial to the entire nation's survival.

Americans are prone to characterizing cancer and rare disease diagnoses as inexplicable, senseless tragedies that strike without warning, and the major thrust of our society's cancer activism is toward curing it, not preventing it. "How did we get to this point?" epidemiologist Devra Davis asks. "Years before any modern industrial nation started an official war on the disease, in the 1930s, researchers in Germany, Japan, Italy, Scotland, Austria, England, Argentina, the United States and France had shown that where people lived and worked affected their chances of getting cancer." We reached this point because other concerns have been privileged above the health of the citizenry and our right to knowledge that would protect our health. Those groups that profit or find advantage when ordinary people fail to make the connection between toxic exposure and disease have exerted great effort to distract us. We affix ribbon-shaped magnets to our cars to support a war on cancer, yet there are no ribbons honoring the fight against the insidious violence of toxic exposure, specifically that of industrial and military pollution.[3]

As citizens sought explanations for this violence, their questions revealed fundamental paradoxes at the heart of the national security state. Citizen stories, corroborated by journalists, epidemiologists, and government documents, demonstrate that for the latter half of the twentieth century the U.S. government violated basic democratic principles and knowingly contaminated the food supply, the air, and the water its citizens relied on under the pretense of keeping Americans safe and of preserving democracy and American dominance in the global order. In order to bring Americans on board for the nuclear arms race, the U.S. government secretly paid a supposedly objective science journalist to shape American thinking about the bomb and educate the nation on the potential benefits (many of them mythical) of nuclear technology. Representatives of the government conducted medical research on the bodies of citizens without obtaining informed consent or warning them of the dangers posed in the areas in which they were living, working, and raising their children. Federal employees not only blatantly lied to citizens who pinpointed radiation-related livestock deaths but also told them their own ineptitude had caused the deaths. Government lawyers repeated these lies in courts of law, intimidated witnesses, and effectively thwarted the justice system. Thousands of citizens—among them children, mothers, breadwinners, and veterans—died of radiation-related illnesses, becoming casualties of the Cold War.

Because they have made the connection between toxic exposure and cancer, many in regions downwind of nuclear and uranium sites are reluctant to trust government and industry representatives who promise nuclear waste can be safely transported through their regions, that nuclear tests can be contained underground, that nuclear power plants can produce clean energy, and that in situ leach mining can extract uranium for those power plants without endangering their children. Those people who retell the stories of the past advocate for future generations with humble effectiveness. Their stories have helped precipitate epidemiological research, compensation legislation, test bans, and toxic site cleanup; have shut down the uranium industry on Indian reservations across the American Southwest; have blocked major military operations, including the Divine Strake test; and have stonewalled efforts to site nuclear waste at Yucca Mountain, adjacent to the Nevada Test Site.

Despite such successes—which most consider avoided tragedies rather than "victories"—radiation-affected people see no end to their exhausting work. They understand all too well that there is money in nuclear technology. Test bans can be overturned, politicians come and go, scientific studies lose funding and are "misfiled," and national security crises prompting new calls for nuclear weaponry and domestic energy production arise.

The year 2011 illustrated the vulnerability of nuclear technology to natural disasters around the globe. In the space of three months, nuclear facilities were threatened by earthquake-induced tsunami waves (Fukushima Daiichi Nuclear Power Plant in Japan), wildfire (Los Alamos National Laboratory in New Mexico), and flooding (Cooper and Fort Calhoun Nuclear Stations in Nebraska). While the residents of New Mexico and Nebraska appear to have been spared serious exposure in this round of disasters, the Fukushima Daiichi plant experienced three meltdowns and hydrogen explosions and spilled prodigious quantities of radioactive coolant water into the ocean. Thousands of Japanese citizens were evacuated, and Arnold Gunderson, a former nuclear power industry executive, predicted the Fukushima disaster would rank as "the biggest industrial catastrophe in the history of mankind." Despite its severity, the Fukushima crisis failed to hold Americans' interest for long, despite the fact that twenty-three American nuclear power plants operate using General Electric boiling-water reactors with Mark I containment systems identical to those that failed at Fukushima. At the time of this writing, in March 2014, Fukushima is still many years away from being safely decommissioned. Rates of thyroid cancer are reported to be increasing in the Fukushima prefecture, and tsunami debris from Fukushima still washes up occasionally on the Pacific coast of the

United States, reminding the citizens who walk the beaches that the world is indeed a small place and filled with forces beyond our control. While much can be done to protect citizens from the nuclear dangers already in existence, no technological advancements can ever make nuclear power or weaponry a safe bedfellow.[4]

Downwinders and uranium-affected people understand that long after the mushroom clouds have dissipated and the yellow monster has gone quiet and the reporters have moved on to the next big story, radiological contamination stays, and people must continue with their ordinary lives. What matters most in the long run is not technological advances, compensation money, a big padlock on the Nevada Test Site, or a grand international declaration that uranium is indeed a monster; what matters, is the stories. Only the stories can rival radiation for longevity, as long as people keep telling them. Bearing witness may seem futile, but it is how we make sense of our losses and our lives and the structures of power that shape society and history. Radiation-affected people understand that if they stop sharing their experiences, there will be no one to rally when the next crisis arises and the media spotlight returns.

In studying the multiple narratives of domestic nuclear testing and uranium extraction, it becomes clear that the validity of the scientific record, the official historical record, and the unofficial vernacular record are all in question. They compete with each other, but to some degree, they are also entangled in each other; thus each must be examined in light of the insight offered by other records. It is tempting to romanticize citizen memory, with its ring of pathos, ordinariness, and capacity for colorful language and exaggeration, in contrast to official memory, which is often characterized as cold, formal, and accurate. Yet this "neat binary opposition" obscures a great deal, Alon Confino reminds us, for "in the real world, things are not as neat. Not only is vernacular memory not as saintly and official memory not as brutal, but they constantly commingle." Both records, historian Richard White points out, "tell only some stories." While resisting the temptation to establish a strict dichotomy between vernacular and official records, it is essential to note the disparity in their representation and power. The vernacular record may captivate the imagination and, periodically, the press, yet the official record—and the complicated laws protecting government and industry in which that official record is tangled—ultimately hold sway both in historical analyses and legal outcomes. The work of the engaged citizen and the engaged scholar, then, lies in traversing the places where the official, the vernacular, the national, and the local collide, conflict, and collude.[5]

Near the end of our conversation in January 2008, Ivan Sidney reflected on rumors of the resumption of weapons testing. "I hear about what's happening . . . and in the beginning, [I thought] I don't know how you can beat this so-called great government, when all of it is progress. Because people want electricity, people want bombs, I guess . . . [people] want to go to space. Well, what's good up there, you know?" Turning a broken metal piece from a local well over in his hands as my tape recorder hummed quietly on his desk, Sidney told me, "We need things here. I have my people that have a need." Prior to our interview, he had spent the morning trying to track down a replacement piece for the well, which numerous residents of Polacca relied on for water. "I hear about government still wanting to do things with uranium. They know what it's causing," he mused. "And you have these nuclear power stations where supposedly they're fault free. We've seen where they went down. And now they're gonna be hauling some of these spent rods on the main highway. I don't know, what can we do?"

Behind him, out the window of his office, a winter storm gathered around the homes perched on First Mesa, the five-hundred-year-old Hopi village where he was raised. Sidney paused, then answered his own question. "I am willing to testify."

NOTES

1. LIVING UNDER THE CLOUD

1. C. Peterson interview. All subsequent quotations regarding C. Peterson are from this source unless cited otherwise.

2. The Nevada National Security Site has been known by many names over the decades. Most residents in the region continue to refer to it as the Nevada Test Site or test site so this book does so as well.

3. Judy Pasternak discusses cultural and spiritual beliefs against mining in *Yellow Dirt*. George Tutt, interview in Navajo by Timothy Benally, Shiprock NM, December 1995, and translated by Esther Yazzie-Lewis and Timothy Benally, in Brugge, Benally, and Yazzie-Lewis, *Navajo People*, 15. All subsequent quotations from Tutt are from this source unless cited otherwise.

4. Esther Yazzie-Lewis and Jim Zion, "Leetso, the Powerful Yellow Monster," in Brugge, Benally, and Yazzie-Lewis, *Navajo People*, 2.

5. Code-named Fat Man and Little Boy, the nuclear weapons dropped on Hiroshima and Nagasaki killed thousands immediately and left behind a massive blanket of radioactive fallout. Medical studies by the Atomic Bomb Casualty Commission (later renamed the Radiation Effects Research Foundation) documented genetic mutations that led to cancer, birth defects, and other health problems. Richard Rhodes notes that in Hiroshima alone, "five-year deaths related to the bombing reached 200,000." In Nagasaki, 70,000 had

died by the end of 1945, and the five-year death toll related to the bombing reached 140,000. See Rhodes, *Making of the Atomic Bomb*, 734, 740. Truman is quoted in Sharp, *Savage Perils*, 121; and Groves is quoted in Sharp, *Savage Perils*, 133.

6. G. Nisson interview. All subsequent quotations from G. Nisson are from this interview unless cited otherwise.

7. Sharp, *Savage Perils*, 153.

8. "The 36-Hour War," *Life*, 34, 32.

9. Solnit, *Savage Dreams*, 6.

10. Federation of American Scientists, "Plot of Worldwide Nuclear Tests"; Johnston, "Half-Lives," in *Half-Lives*, 4; Nuclear Weapon Archive, "Gallery"; and Dawson and Madsen, "Uranium Mine Workers," in Johnston, *Half-Lives*, 118.

11. Johnston, "Half-Lives," in Johnston, *Half-Lives*, 9.

12. Dr. John Gofman, interview by Carole Gallagher, San Francisco, March 1992, in Gallagher, *American Ground Zero*, 329; Sternglass, "Cancer"; Clarfield and Wiecek, *Nuclear America*, 359; and Dyson, "Comments," 27.

13. National Cancer Institute, "Estimated Exposures."

14. See Lindee, *Suffering*; Advisory Committee on Human Radiation Experiments, *Interim Report*; and Johnston, "'More Like Us,'" in *Half-Lives*, 25–54.

15. Evans, "Medical Uses," 68.

16. Judge Bruce S. Jenkins, Memorandum Opinion, 10 May 1984, 200, 205, 211, 316, in *Irene Allen et al. v. United States*; and Duncan A. Holaday, Wilfred D. David, and Henry N. Doyle, "An Interim Report of a Health Study of the Uranium Mines and Mills" (Washington DC: Federal Security Agency, Public Health Service, Division of Occupational Health, and Colorado State Department of Public Health, May 1952), reprinted in Eichstaedt, *If You Poison Us*, 203–17.

17. Simon, Bouville, and Land, "Fallout," 7, 1.

18. Fischer, *Citizens*, 49.

19. Yazzie-Lewis and Zion, "Leetso," 2.

20. Richard Dorson, "Oral Tradition and Written History: The Case for the United States," *Journal of the Folklore Institute* 1 (1964): 233–34, quoted in Allen, "Personal Point of View," 112; and Schrager, "Stories Communities Tell," 220.

21. Kansteiner, "Finding," 184; and Burke, "History," 98.

22. Geertz, *After the Fact*, 2–3.

1. Tommy James, interview in Navajo by Phil Harrison, Cove AZ, December 1995, and translated by Esther Yazzie-Lewis and Timothy Benally, in Brugge, Benally, and Yazzie-Lewis, *Navajo People*, 124. All subsequent quotes from James are from this source unless cited otherwise.

2. Peter Iverson's *Diné* is an exemplary reference on Navajo history.

3. More than forty thousand Native Americans served in the armed forces during World War II, and Selective Service data indicates that the vast majority of eligible Native American men registered for service. For additional background on the war's impact on Native American communities, see Bernstein, *American Indians*.

4. Richard Rhodes provides an extensive history of the Manhattan Project and the development of the bomb in *The Making of the Atomic Bomb*.

5. Eichstaedt, *If You Poison Us*, 11; Ringholz, *Uranium Frenzy*, 5–6; and Benally, "Navajo Uranium Miners."

6. Ringholz, *Uranium Frenzy*, 5–6.

7. Yazzie-Lewis and Zion, "Leetso," 2; and Eichstaedt, *If You Poison Us*, 26.

8. Eichstaedt, *If You Poison Us*, 28. Harry Goulding described his role in the uranium boom in "Navajos Hunt Big Game," 238.

9. Eichstaedt, *If You Poison Us*, 27.

10. Eichstaedt, *If You Poison Us*, 37, 40.

11. See Clark, *Lone Wolf v. Hitchcock*; and Anaya, "International Law," in Grounds, Tinker, and Wilkins, *Native Voices*, 155–85. While some scholars, activists, and politicians—indigenous and nonindigenous—have argued that BIA negligence and the importance of Native sovereignty necessitate abolition of the agency, I disagree. Like all government agencies, the BIA is flawed but still provides crucial services to Native American peoples, health care especially. Until Native communities are able to establish equivalent or superior services and legal protections on their own, the BIA, imperfect as it may be, remains necessary. The tragedy of the termination movement of the 1940s and 1950s illustrates this point (see chapter 6 of this volume).

12. Pino interview. All subsequent quotes from Pino are from this source unless cited otherwise.

13. For background on the emergence of the Atomic Age and discussions of radiation in popular culture, see Boyer, *By the Bomb's Early Light*.

14. Kuletz, *Tainted Desert*, 25; Johnston and Dawson, "Resource Use," in Johnston, *Who Pays the Price?*, 143; Miller, "Changes," 438; and Choudhary, "Navajo Nation," 10, 13.

15. Waconda, testimony, World Uranium Hearings. All subsequent quotes from Waconda are from this source unless cited otherwise.

16. Judy Pasternak found Goulding's statement in an unpublished oral history interview. See Pasternak, *Yellow Dirt*, 46.

17. Benally interview. All subsequent quotes from Benally are from this source unless cited otherwise.

18. Toelken, *Anguish of Snails*, 42.

19. Toelken, *Dynamics of Folklore*, 276–77.

20. Wilbert L. Dare, *Uranium Mining in the Lukachukai Mountains, Apache County, Arizona, Kerr-McGee Oil Industries, Inc.* (Washington DC: U.S. Bureau of Mines, 1961), 9, quoted in Eichstaedt, *If You Poison Us*, 44. There is ample literature discussing stereotypes about Native Americans. See, for example, Bordewich, *Killing the White Man's Indian*; and Deloria, *Indians in Unexpected Places*.

21. Colleen O'Neill's *Working the Navajo Way* is an excellent resource on the history and dynamics of the Navajos' engagement with wage labor.

22. Doug Brugge and Rob Goble, "The History of Uranium Mining and the Navajo People," *American Journal of Public Health* 92 (September 2002): 1410–19, reprinted in Brugge, Benally, and Yazzie-Lewis, *Navajo People*, 25–48, with "no scientific" quote on 30 and "radon in mines" quote on 32; and "cooperating" quote from Public Health Service report in Eichstaedt, *If You Poison Us*, 206.

23. Archer quoted by Ball, *Cancer Factories*, 46; and Dawson, "Navajo Uranium Workers," 392.

24. Brugge and Goble, "History of Uranium Mining," in Brugge, Benally, and Yazzie-Lewis, *Navajo People*, 34; and Eichstaedt, *If You Poison Us*, 203.

25. Brugge and Goble, "History of Uranium Mining," 32; and Lorraine Jack, interview in Navajo by Phil Harrison, Cudei NM, December 1995, and translated by Esther Yazzie-Lewis and Timothy Benally, in Brugge, Benally, and Yazzie-Lewis, *Navajo People*, 52–53. All subsequent quotes from Lorraine Jack are from this source unless cited otherwise.

26. Eichstaedt, *If You Poison Us*, 208, 49.

27. Eichstaedt, *If You Poison Us*, 211.

28. Zang interview. All subsequent quotations from Zang are from this source unless cited otherwise.

29. Logan Pete, interview by Phil Harrison, Mitten Rock NM, in Brugge, Benally, and Harrison, *Memories Come to Us*, 6.

30. Eichstaedt, *If You Poison Us*, 207; and Dawson, "Navajo Uranium Workers," 392–93.

31. Eichstaedt, *If You Poison Us*, 211.

32. George Lapahie, interview in Navajo by Timothy Benally, Two Grey Hills NM, December 1995, and translated by Esther Yazzie-Lewis and Timothy Benally, in Brugge, Benally, and Yazzie-Lewis, *Navajo People*, 80; and Harrison interview. All subsequent quotations from Harrison are from this source unless cited otherwise.

33. Dawson, "Navajo Uranium Workers," 393. Earl Saltwater Jr. quoted in Helms, "Red Valley"; and Dawson, "Navajo Uranium Workers," 393.

34. See Shannon and Graef, "Lead Intoxication," 380–82; and Moorhead, "Milking It." The question of radiation contaminating formula made the news as recently as 2011, following the accident at the Fukushima Daiichi nuclear power plant in Japan. See Knox, "How Risky?"

35. Donald Yellowhorse, interview in Navajo by Phil Harrison, Cove AZ, and translated by Timothy Benally, in Brugge, Benally, and Harrison, *Memories*, 12.

36. Robinson, *Uranium Mill*, 4; and Shuey and Ronca-Battista, *Report*, 4.

37. North Country Healthcare lists the ailments the federal government has acknowledged are connected to transportation of ore on its web page "Radiation Exposure."

38. Makhijani and Saleska, "Production of Nuclear Weapons," in Makhijani, Hu, and Yih, *Nuclear Wastelands*, 33; Eichstaedt, *If You Poison Us*, 212; and Dawson, "Navajo Uranium Workers," 393.

39. Eichstaedt, *If You Poison Us*, 212.

40. Nelkin, "Native Americans," 2; Makhijani and Saleska, "Production," 33–34, 35; and Yih et al., "Uranium Mining," in Makhijani, Hu, and Yih, *Nuclear Wastelands*.

41. Eichstaedt, *If You Poison Us*, 140, 141.

42. Pasternak, "They Took Shelter," 5; Joseph M. Hans, Jr., and Richard L. Douglas, *Radiation Survey of Dwellings in Cane Valley, Arizona, for Use of Uranium Mill Tailings* (Las Vegas: U.S. Environmental Protection Agency, Office of Radiation Program, August 1975), quoted in Eichstaedt, *If You Poison Us*, 142; Ford, Bacon, and Davis Utah, Inc., *Engineering Assessment of Inactive Uranium Mill Tailings, Monument Valley Site, Monument Valley, Arizona*, report for U.S. Energy Research and Development Administration, Grand Junction CO, 31 March 1977, quoted in Eichstaedt, *If You Poison Us*, 142–43.

43. Pasternak, "They Took Shelter," 1, 3; and Kramer, "Spokane Tribe."

44. Yih et al., "Uranium Mining," 121.

45. Brugge, deLemos, and Bui, "Sequoyah Corporation"; Nelkin, "Native Americans," 7; and Rita Begay quoted in Kuletz, *Tainted Desert*, 26.

46. Robinson also notes livestock growers had trouble selling their meat after the spill. Paul Robinson, "Uranium Production," in Bryant and Mohai, *Race and the Incidence*, 158.

3. HOME ON THE RANGE

1. Annie Corry, interview by Carole Gallagher, Cedar City UT, August 1988, in Gallagher, *American Ground Zero*, 272. All subsequent quotes from Corry are from this source unless cited otherwise.

2. Laurence's description of the Trinity test is reprinted in Titus, *Bombs in the Backyard*, 17. Sharp, *Savage Perils*, 129. Journalist David Goodman discussed William Laurence's relationship with the AEC in detail on the 5 August 2005 *Democracy Now* radio program. See Goodman and Goodman, "Hiroshima Cover-up."

3. For background on the Pacific Proving Ground and its effect on the Marshall Islanders, see Embassy of the Republic of the Marshall Islands, "Nuclear Issues"; and Dibblin, *A Day of Two Suns*. Gordon Dean quoted in "You Go to Desert Rock," *Armed Forces Talk* 420 (19 September 1952): 2, located in folder 17, box 486, Gunn McKay Papers, Mss 86, Special Collections, Utah State University, Logan (hereafter McKay Papers). For a thorough discussion of the selection of the NTS site, see Schoemehl, "Nuclear Reactions," 21–75.

4. See Beck, Nieves Zedeño, and Furlow, "Time, Trust"; and Solnit, *Savage Dreams*, 30. The inception of testing in Nevada did not put an end to Pacific testing. Concerns about exposure to downwind residents in Utah after the 1953 nuclear test Harry shifted American nuclear tests back to the Pacific Proving Ground; one of these tests inadvertently made Nevada the test site of choice again. The 1954 Bravo shot, a thermonuclear test conducted in 1954, exploded with 15 megatons of force, equivalent to 15 million tons of TNT, or almost double what the AEC had predicted. A shift in expected wind patterns carried the massive plume of ensuing fallout over Marshall Islanders, U.S. servicemen, and a crew of twenty-three Japanese fishermen on fishing vessel *Daigo Fukuryu Maru* (Lucky Dragon 5) farther out in the Pacific. The fishermen and their catch were heavily contaminated, leading to an uproar in Japan and a $2 million compensation payment from the

United States. Two years later, Japanese film director Ishirō Honda paid homage to the crew of the fishing boat in a scene in *Godzilla, King of the Monsters* (1956), which told the story of a massive underwater beast whose hibernation was disturbed by nuclear testing. See Internet Movie Database (IMDB), "*Godzilla.*" While the Pacific Proving Ground would be utilized for later tests, the future of the Nevada Test Site had been assured for the time being, as it offered the AEC a location that could be relied upon not to inspire any more international stories about radioactive monsters emerging from the oceans. If strange stories did arise downwind, the AEC felt confident it could keep them under wraps.

5. Memorandum, W. S. Parsons to J. E. Hull, 12 May 1948, quoted in Schoemehl, "Nuclear Reactions" 37, 76n32, 154. Glazer and Glazer, "National Security," 271.

6. Bulloch interview. All subsequent quotes from Bulloch are from this interview unless otherwise indicated.

7. Miller, *U.S. Atlas of Nuclear Fallout*, 133–84.

8. See the *Iron County (Cedar City UT) Record*: "Unusual Gift for First Lady," 9 July 1953; "Wool Growers Auxiliary Announces 'Make It Yourself with Wool' Contest," 5 March 1953; and "Pot-Luck Supped [*sic*] Held by Members of Wool Growers Aux.," 30 July 1953. For examples of additional coverage of the ranching industry, see the following 1953 issues of the *Iron County (Cedar City UT) Record*: 28 May, 18 June, 20 August, 27 August, and 17 September. Mrs. McRae Bulloch and Mrs. Kern Bulloch are Lillian and Vera, the wives of ranchers McRae and Kern Bulloch, respectively.

9. Stephen Brower, *Annual Report of Extension Service, 1953, County Extension Agents—Iron County, State of Utah, from November 1, 1952, to October 31, 1953*, 12, in 1953 folder, box 29, University Extension Service Reports, 19.1/1:47, in Special Collections, Utah State University, Logan.

10. Seegmiller interview.

11. For an extensive analysis on the impacts of livestock grazing on rangeland in the American West, see Fleischner, "Ecological Costs," 629–44. See also Colorado Plateau–Land Use, "Cattle and Sheep Grazing History."

12. Jones, "Regionalization," 111.

13. Daniel Robert Sheahan, quoted in "Reporter's Transcript of Portions of Testimony of Daniel Robert Sheahan, Daniel Sheahan, and Joe B. Sanders," 11, 12, and 13 September 1956, in Bulloch et al. vs. United States, reprinted by of U.S. House, *Health Effects*, 2:1088.

14. U.S. House, *Health Effects*, 2:1089, 1093.

15. U.S. House, *Health Effects*, 2:1096, 1094–95.
16. Utah State Agricultural College, a land grant institution located in the northern Utah city of Logan, opened its doors in September 1890. It became Utah State University in March 1957. See Parson, "Brief History." Stephen Brower, interview by Carole Gallagher, Provo UT, February 1988, in Gallagher, *American Ground Zero*, 263, 264.
17. Melvin quoted in Hacker, *Elements of Controversy*, 107. See also Hacker, "'Hotter Than a $2 Pistol,'" in Hevly and Findlay, *The Atomic West*, 160.
18. For "serious . . . problem" quote, see "Minutes," AEC meeting no. 866, 22 May 1953, U.S. House, *Health Effects*, 1:146. For Utah media responses to the alarming atomic events of 1953 (in addition to the sheep deaths), see "Atomic Dust Discovered on Cars near St. George as AEC Sets Roadblock," *Washington County (St. George UT) News*, 30 April 1953; and the following front-page stories in *Deseret (Salt Lake City) News*: "St. George Goes Indoors as Atomic Cloud Passes," 19 May 1953; "Excess Radioactivity Delays 8th A-Blast," 1 May 1953; "A-Cloud Moving over Utah May Pass over S.L.," 8 May 1953; "Atomic Blast Shakes Town 200 Miles Off," 19 May 1953; "AEC Sends Experts to Check Sick Utahns," 20 May 1953; and "AEC Denies Blast Caused Utah Illness," 21 May 1953. The *Deseret (Salt Lake City) News*'s coverage of activities at the test site greatly exceeded the coverage in smaller, local southern newspapers such as *Iron County (Cedar City UT) Record* and *Washington County (St. George UT) News*. This trend is discussed in chapter 5. The Zuckert quote is from "Minutes," AEC Meeting no. 866, 22 May 1953, in U.S. House, *Health Effects*, 1:146.
19. Headlines are from the front pages of *Deseret (Salt Lake City) News*, 14 and 29 May 1953.
20. Stephen L. Brower, letter to Governor Scott Matheson, 14 February 1979. Reprinted in U.S. House, *Health Effects*, 1:567.
21. Maj. R. J. Veenstra, V. C., report to Mr. Allare, Atomic Energy Commission, 17 June 1953, reprinted in U.S. House, *Health Effects*, 1:536. Extensive documentation of the various aspects of the AEC's investigation into the sheep deaths is contained in U.S. House, *Health Effects*, vol. 2. A table of contents for all documents is provided on 1424–28.
22. Ball, *Justice Downwind*, 205.
23. "Cooperative Contract, No. AT(11-1)-80, between Utah State Agricultural College and the U.S. Atomic Energy Commission for Atomic Energy Research and Development," 10 June 1950, folder 2, box 114, Utah Agricultural Experiment Station Director's Files, RG 18:17, Special Collections, Utah State

University, Logan (hereafter ESDF); and letter, R. H. Walker, Dean, to Dr. Walter J. Kowalski, Research Contracts Division, 5 April 1951, folder 2, box 114, ESDF.

24. See Utah State University, "About Us."

25. Willard "Bill" Peterson interview. All subsequent quotations from B. Peterson are from this source unless cited otherwise. Marjorie Black, interview by Carole Gallagher, Delta UT, November 1988, in Gallagher, *American Ground Zero*, 279–80.

26. Black interview in Gallagher, *American Ground Zero*, 281.

27. Stephen L. Brower, to Committees Investigating Damage from Fallout in Southern Utah, Statement prepared for *Hearing on Health Effects of Low-Level Radiation*, 19 April 1979, reprinted in U.S. House, *Health Effects*, 1:235.

28. Historian Barton Hacker provides an excellent survey of the official investigation in "Hotter Than a 2$ Pistol," in Hevly and Findlay, *Atomic West*, 157–75. For an extended discussion, see Hacker, *Elements of Controversy*, 106–30. See also the following articles in the *Iron County (Cedar City UT) Record*: "'Malnutrition' Diagnosis Resented: Authoritative Report Expected Soon," 2 July 1953; "'Malnutrition' Explains Sheep Mystery Disease," 25 June 1953; "Cause of Death: Radiation or Malnutrition?," 16 July 1953; and "Investigation Goes Forward to Solve Sheep Mystery Death," 13 August 1953.

29. Stephen Brower, letter to Governor Scott Matheson, 14 February 1979, reprinted in U.S. House, *Health Effects*, 1:568. Richard Miller offers an account of the confrontation drawn from an interview with Clark's son, Ken, who spoke with his father shortly after his run-in with Rust. See *Under the Cloud*, 184–85. The details in Miller's history vary slightly from Brower's account. I have offered Brower's description because it is the only version of the incident that I know of that was authored by an actual witness to the confrontation.

30. Delayne Evans, interview by Carole Gallagher, Parowan UT, October 1988, in Gallagher, *American Ground Zero*, 276. All subsequent quotations from Evans are from this source unless cited otherwise.

31. "A city councilman . . ." quote in interview with McRae Bulloch by Carole Gallagher, March 1986, in Gallagher, *American Ground Zero*, 268. All other quotes from Bulloch in this paragraph from interview by author.

32. L. Johnson interview. All subsequent quotations from Johnson are from this source unless cited otherwise.

33. U.S. Atomic Energy Commission Report, 6 January 1954, excerpted in U.S. House, *Health Effects*, 1:571.

34. Minutes, Meeting of Livestockmen and AEC Officials, 13 January 1954, Fire House Conference Room, Cedar City UT, 3, folder 2, box 486, McKay Papers. All subsequent quotes from this meeting are from this source unless cited otherwise.

35. Geertz, *Local Knowledge*, 75–76. Fischer's *Citizens, Experts* is also an excellent reference.

36. Dr. Harold Knapp, "Sheep Deaths in Nevada following the 1953 Nuclear Tests," report prepared for the Senate Subcommittee on Health and Scientific Research, Senate Committee on the Judiciary, House Subcommittee on Oversight and Investigations, and the Governor of Utah, 18 June 1979, in U.S. House, *Health Effects*, 1:288.

37. Progress Report, "Effect of Nutrition on Pathology and Productivity of Range Sheep," conducted by USAC Agricultural Experiment Station in cooperation with Atomic Energy Commission, 62a, in folder 1, box 82, ESDF. See also the *Iron County (Cedar City UT) Record*: "USAC Receives Grant to Continue Sheep Loss Survey," 18 March 1954; and "$25,000 Grant by Atomic Energy Commission Made to Study Sheep Malady in South Utah,"19 August 1954.

38. "Complaint," 17 February 1955, *Bulloch et al.* Reprinted in U.S. House, *Health Effects*, 1:630. See also the *Iron County (Cedar City UT) Record*: "Livestock Owners Consider Suit to Recover Herd Loss," 11 February 1954; and "Local Sheep Raisers File Suit with Govt. for Loss," 3 March 1955.

39. Rosenberg, "Sorrow and the Pity," in Hensley, *Boundaries of Freedom*, 23.

40. Author's transcription of U.S. Atomic Energy Commission, U.S. Air Force, and Lookout Mountain, *Atomic Tests in Nevada*.

41. Stegner, *Mormon Country*, 98. A notable exception to this trend was the LDS church's decision to oppose the development of the MX missile system in the Great Basin in the early 1980s. See Chip Ward, "MX Marks the Spot," in *Canaries on the Rim*, 51–61.

42. Darlene Phillips, interview by Carole Gallagher, Bountiful UT, December 1991, in Gallagher, *American Ground Zero*, 303, 304.

43. Schoemehl, "Nuclear Reactions," 79; and Michelle Thomas, telephone conversation with author, 13 January 2006.

44. Dan Bushnell, in "Reporter's transcript of portions of closing argument of attorney Dan S. Bushnell and comments of the court," 1 and 2 October 1956, *Bulloch et al.*, in U.S. House, *Health Effects*, 2:1018; and Bulloch interview.

45. Judge Sherman Christensen, "Memorandum, Findings of Fact and Conclusions of Law," Retrial, *Bulloch et al.*, 4 August 1982, 23; and Judge Sherman

Christensen, "Opinion," 26 October 1956, 1, *Bulloch et al.* Both are available online via a search at http://campus.westlaw.com (accessed 29 April 2005).

46. S. Johnson interview. All subsequent quotations from S. Johnson are from this source unless cited otherwise.

47. Author's transcription of U.S. Atomic Energy Commission, U.S. Air Force, and Lookout Mountain, *Atomic Tests in Nevada.*

48. Claudia Peterson, interview by Carole Gallagher, St. George UT, October 1988, in Gallagher, *American Ground Zero,* 125.

49. While Peterson was unable to verify the year of this die-off conclusively, he recalled it in exceptional detail and pinpointed it as having occurred in late November and most likely 1954. List and Machta's report quoted in Miller, *Under the Cloud,* 247–48, 511n33.

50. Chip Ward, "Cowboys in Gas Masks Find a Damn Good Place to Dump Used Razor Blades," in *Canaries on the Rim,* 99, 101.

51. Lee Davidson and Joe Bauman broke the story about the sheep deaths in their article "Toxic Utah: A Land Littered with Poisons," *Deseret (Salt Lake City) News,* 28 February 2001.

52. Dawson, Charley, and Harrison, "Advocacy and Social Action," in Brugge, Benally, and Yazzie-Lewis, *Navajo People,* 69.

53. Pasternak, "Oases in Navajo Desert," 1–2.

54. Gilbert Fraga, interview by Carole Gallagher, August 1986, Sacramento, in Gallagher, *American Ground Zero,* 35.

4. LOCALLY GROWN

1. Maiser et al., "Locavores."

2. Rollin, *Farm Animal Welfare,* 8–9.

3. Michael Pollan discusses breakthroughs in understanding and manipulating soil fertility in *Omnivore's Dilemma,* 146–47. Rachel Carson addresses the evolution of pesticides from chemical warfare research in *Silent Spring.* Dan McNichol describes the development of the interstate highway system in *The Roads that Built America.* In *Thermo King Model C Transport Refrigeration Unit,* the American Society of Mechanical Engineers International and Thermo King Corporation offer a brief history of changes in refrigerated transport.

4. Pollan, *Omnivore's Dilemma,* 34.

5. Rollin, *Farm Animal Welfare,* 8–9. Prison statistic is from Liptak, "Inmate Count"; and farm statistic and quote are from Pollan, *Omnivore's Dilemma,* 34.

6. Dickson interview by author. All subsequent Dickson quotations are from this source unless cited otherwise.

7. Kay Millet, interview by Carole Gallagher, near Cedar City UT, March 1988, in Gallagher, *American Ground Zero*, 146. All subsequent quotations from Millet are from this source unless cited otherwise.

8. David Arthur Timothy, deposition, 18 January 1981, 51, 53–54, in David Timothy et al. v. United States, case no. C-80-0445 A (hereafter Timothy deposition), in boxes 1 and 2, acc. 021-91-0024, loc. 921254-921256, National Archives and Records Center, Denver (NARC–Denver).

9. Timothy deposition, 55, 57, 59, 60. See Robert Pendleton, deposition, 16 January 1981, in *Dave Timothy et al.* (hereafter Pendleton deposition), in boxes 1 and 2, acc. 021-91-0024, loc. 921254-921256, NARC–Denver.

10. Sidney interview. All subsequent quotations from Sidney are from this source unless cited otherwise.

11. Timothy deposition, 63.

12. Nuclear Weapon Archive, "Operation Upshot-Knothole"; and author's transcription of U.S. Atomic Energy Commission, U.S. Air Force, and Lookout Mountain, *Atomic Tests in Nevada*.

13. Q. Nisson interview. All subsequent quotations from Q. Nisson are from this interview unless cited otherwise.

14. Frank Butrico, interview in documentary *The Cover-Up at Ground Zero*. Quote is from author's transcription after viewing and tape-recording video of the documentary in private collection of Quentin Nisson, Washington City UT, 21 October 2005.

15. "Chamber of Commerce Hear Discussions, See A-Bomb Fallout Picture," *Washington County (St. George UT) News*, 11 April 1955. St. George resident Nora Lyman commented enthusiastically on the film in her *Washington County (St. George UT) News* column, Observation; and Elmer Pickett, interview by Carole Gallagher, St. George UT, February 1984, in Gallagher, *American Ground Zero*, 150. The ABC documentary *Cover-Up at Ground Zero* lists local film participants Earl Cox, Tony Reynolds, and Roy Renoff as having succumbed to cancer. Sheldon Johnson also recalled in an interview with this author that Cal Cottam, the service station attendant featured in the film, died of cancer.

16. Schoemehl's treatment of the zone management program, based on extensive archival research, is the most thorough examination of local interactions

between the AEC and downwind residents to date. See "Nuclear Reactions," 190, 194, 210.

17. See *Washington County (St. George UT) News*, 3 March 1955, B1, B3. Michelle Thomas pointed out the original location of the Rugged West on a driving tour of St. George during a 19 October 2005 interview by author. The motel no longer exists.

18. Timothy deposition, 18–19, 20, 84, 85–87.

19. Timothy deposition, 45, 46–49.

20. Timothy deposition, 64, 65, 69. For figures and descriptions illustrating the radiation-measuring process Susan Fisher and others underwent, see Pendleton et al., "Differential Accumulation," 1254.

21. Timothy deposition, 65; and Timothy interview.

22. Robert C. Pendleton, PhD, Curriculum Vitae (exhibit 1), Pendleton deposition. Statements about Pendleton's work history are discerned from Pendleton's curriculum vitae, his testimony in the Timothy case, and the abstracts of his articles. He discusses setting up the network in Pendleton et al., "Differential Accumulation," 1253; and in Lloyd et al., "Comparison," 427. From Pendleton's curriculum vitae, it appears the urgency of the iodine-131 (I-131) research displaced the Cs-137 research; while his first article on I-131 appears in 1963, his research on Cs-137 and humans does not appear until 1966.

23. Pendleton discusses this trip in "Differential Accumulation," 1253; and quoted material in Pendleton deposition, 57–58. See also Miller, *U.S. Atlas of Nuclear Fallout*, 329–40.

24. Pendleton et al., "Differential Accumulation," 1253.

25. Pendleton et al., "Differential Accumulation," 1253, 1260, 1253.

26. Kricher, *Ecology of Western Forests*, 45. See Miller's discussion of "rainouts" in *U.S. Atlas of Nuclear Fallout*, 41–42.

27. Schoemehl, "Nuclear Reactions," 347; Pendleton deposition, 66.

28. Chip Ward, "Breakfast Cereal for Two-Headed Babies," in *Canaries on the Rim*, 87.

29. Schoemehl, "Nuclear Reactions," 379.

30. Pendleton deposition, 52, 54.

31. Pendleton deposition, 198, 192; and Timothy interview.

32. For discussion of historic milk consumption, see Schoemehl, "Nuclear Reactions," 375–76; and Dreicer, Bouville, and Wachholz, "Pasture Practice,"

630. Quotes from Gunderson, "National School Lunch Program"; and Yeoman, "Unhappy Meals."

33. Quotes from "Utah Milk Producers Form Promotion Unit," *Deseret (Salt Lake City) News*, 7 May 1953; and a "Drink Milk" advertisement in *Iron County (Cedar City UT) Record*, 12 May 1955.

34. See *Iron County (Cedar City UT) Record*: "Utahns to Celebrate June as 'Dairy Month,'" 28 May 1953; "Milk for the Queen," 11 June 1953; and "Marsha Lewis Wins First Prize in Milk Carton Contest," 11 June 1953.

35. Author's interviews with C. Peterson, G. Nisson, Seegmiller, Dickson, and Thomas.

36. See U.S. AEC, "Off-Site Radiological Safety Plan for Operation Plumbbob," 1 May 1957, quoted in Schoemehl, "Nuclear Reactions," 235. Rulon "Boots" Cox, interview by Carole Gallagher, St. George UT, December 1991, in Gallagher, *American Ground Zero*, 249, 251. Cox's dairy supplied milk to Michelle Thomas's family and many other St. George residents.

37. Quoted in Schoemehl, "Nuclear Reactions," 291.

38. Consumers Union, "Milk All of Us Drink," cover, 109, 111. See the table comparing the strontium content of milk from various cities on 109.

39. Rosen, "Flashback; New Report"; Solnit, *Savage Dreams*, 103–5.

40. Consumers Union, "Milk All of Us Drink," table on 109.

41. Timothy deposition, 91, 94, 96, 95.

42. Agatha Mannering, interview by Carole Gallagher, Ivins UT, April 1984, in Gallagher, *American Ground Zero*, 199.

43. Diane Nielson, interview by Carole Gallagher, Henderson NV, June 1986 and August 1987, in Gallagher, *American Ground Zero*, 157.

44. "Fallout? 'Not Enough to Worry About' Says AEC," *Iron County (Cedar City UT) Record*, 10 March 1955; "Uranium Prospectors Are Informed of AEC Site Tests for Mid-February," *Washington County (St. George UT) News*, 20 January 1955; and U.S. AEC, "Atomic Tests in Nevada" (Washington DC: U.S. Government Printing Office, March 1957), 23, pamphlet in private collection of Preston Jay Truman, Malad ID.

45. Kuletz, *Tainted Desert*, 16.

46. Benally, "Navajo Uranium Miners."

47. Didion, *The White Album*, 65; Colorado Plateau–Land Use, "Modern Climactic Conditions"; and Benally, "Navajo Uranium Miners."

48. Weigle and White offer examples of southwestern water folklore in *The Lore of New Mexico*, 146. See also Toelken, "Traditional Water Narratives,"

191–200. Keith Basso notes that "water imagery . . . pervades the Apache model of wisdom," a concept he discusses throughout *Wisdom Sits in Places*, 159. See also Stegner, *American West as Living Space*, 24.

49. Berry, "Work of Local Culture," 166.

50. Berry, "Work of Local Culture," 166.

5. WRITING DOWN NAMES

1. Thomas interview. All subsequent quotations from M. Thomas are from this source unless cited otherwise.

2. Federal Civil Defense Administration, "Facts about Fallout"(Washington DC: U.S. Government Printing Office, 1955), pamphlet in private collection of Preston Jay Truman, Malad ID.

3. McEnaney, *Civil Defense*, 6.

4. McEnaney, *Civil Defense*, 41, 52.

5. McEnaney, *Civil Defense*, 54.

6. Sharp, *Savage Perils*, 177.

7. McEnaney, *Civil Defense*, 69.

8. Michelle Thomas, "Meditations on My Mother," unpublished essay read during interview by author.

9. The AEC's encouragement of and instructions for shot-watching can be found in 1955 and 1957 AEC pamphlets "Atomic Tests in Nevada," in private collection of Preston Jay Truman, Malad ID, and viewed by author.

10. Schoemehl, "Nuclear Reactions," 3; and AEC, "Atomic Tests in Nevada," 1957, 33.

11. See, "Fallout Aimed, Udall Says," clipping, folder 7, box 485, McKay Papers; and Ina Iverson, interview by Carole Gallagher, Highland UT, October 1988, in Gallagher, *American Ground Zero*, 143.

12. Ena Cooper, interview with Carole Gallagher, St. George UT, March 1988, in Gallagher, *American Ground Zero*, 188; and Martha Bordoli Laird, interview by Carole Gallagher, Carson City NV, November 1988, in Gallagher, *American Ground Zero*, 117–18. All subsequent quotes from these individuals are from these sources unless cited otherwise. At this time and when Butch, Martha's son, was ill, she went by Bordoli. She remarried in the 1970s and went by Laird afterward.

13. Glenna Orton, interview by Carole Gallagher, Parowan UT, January 1984 and December 1991, in Gallagher, *American Ground Zero*, 179. All subsequent quotations from Orton are from this source unless cited otherwise.

14. List and Machta quoted in Miller, *Under the Cloud*, 247. Frederick Schoemehl discusses the Crandalls in "Nuclear Reactions," see especially 222–24.

15. *Tonopah (NV) Times-Bonanza*, 26 October and 16 November 1956 and 24 May 1957.

16. Martha Bordoli Laird's petition is reprinted by Carole Gallagher in Gallagher, *American Ground Zero*, 118.

17. Lewis Strauss, Chairman, AEC, to Mrs. Martha Bordoli, 1957, folder 20, box 486, McKay Papers.

18. Millet interview in Gallagher, *American Ground Zero*, 146. All subsequent quotations from Millet are from this source unless cited otherwise.

19. Helen Nisson interview in documentary *Cover-Up at Ground Zero* with quote from author's transcription. Dorothy See, in "Fallout Aimed, Udall Says," notes Sheldon Nisson's death along with the leukemia death of Jessie Empey's granddaughter (Sheldon's cousin) of St. George in 17 March 1959.

20. Tome quoted in Eichstaedt, *If You Poison Us*, 96–97.

21. C. G. Salisbury, "Cancer Immunity in the Navajo," *Arizona Medicine*, August 1956, 309–10; and Pasternak, *Yellow Dirt*, 136. Eichstaedt verifies the "uranium hill" moniker in *If You Poison Us*, 108; and Stewart Udall refers to the "cancer hill" in "We Aimed for Russia."

22. Robert Carter, interview by Carole Gallagher, Taylorsville UT, October 1988, in Gallagher, *American Ground Zero*, 61–62; Saffer and Kelly, *Countdown Zero*, 248, 257; and U.S. DOE, *Human Radiation Experiments*.

23. Henry Vyner analyzes post-traumatic stress disorder among veterans in "Psychological Effects," 241–61. For discussion of the relationship of narrative to the disorder, see Murphy, "Isolated, Emotional."

24. Ted Przygucki, interview by Carole Gallagher, Las Vegas, July 1986, in Gallagher, *American Ground Zero*, 65–66.

25. Michelle Thomas, phone conversation with author, 25 July 2013.

26. Michelle Thomas, phone conversation with author, 23 February 2006.

27. All quotes from Brown and Mikkelsen, *No Safe Place*, online excerpt chap. 4, p. 1, at http://www.ucpress.edu/books/pages/5524/5524.ch04.html (accessed 14 March 2006).

28. Glazer and Glazer, "National Security," 272–73.

29. Lapahie interview in Brugge, Benally, and Yazzie-Lewis, *Navajo People*, 86.

30. Minnie Tsosie, interview in Navajo by Timothy Benally, Cove AZ, December 1995, and translated by Esther Yazzie-Lewis and Timothy Benally, in Brugge, Benally, and Yazzie-Lewis, *Navajo People*, 158–59.

31. Evan Cooper, interview by Carole Gallagher, March 1988, St. George UT, in Gallagher, *American Ground Zero*, 188.

32. Peterson interview in Gallagher, *American Ground Zero*, with mention of lambs on 125, Geiger counter on 126, and deaths of friends on 125.

33. Elmer Pickett interview in documentary *Cover-Up at Ground Zero*, with quote from author's transcription.

34. In the 1980s, researcher Richard Miller assembled as much data on radiation releases and meteorological patterns as he could gather and reconstructed rudimentary maps of local and national fallout exposure in a two-volume atlas of fallout dispersal from domestic testing. Many, such as Robert Seidel, have criticized Miller's maps, suggesting they indicate "little about actual exposure." While it is true that the full extent of radiological pollution is impossible to apprehend, Miller's maps were the first publicly available representation of fallout dispersal based on the available data. They have been powerful tools for downwinder activists. See Miller, *U.S. Atlas of Nuclear Fallout*, 1:7–33, for a useful discussion of his attempts to reconstruct fallout exposure. See also Miller's *Under the Cloud*; and Seidel, "Books on the Bomb," 534.

35. Griggs, "'Exposed's' Requiem."

36. Authors of a number of professional epidemiological studies have concluded that rates of particular diseases, including thyroid cancer, childhood leukemia, and lung cancer, have increased in downwind areas. While all articles have been vetted by peer review, they remain controversial. Even if these studies stood unchallenged, it is impossible to prove one person's illness was caused by radiation, just as there is no way to prove a lung cancer victim's illness was brought on by their smoking habit. These topics will be discussed in further detail in chapter 6.

37. Claudia Peterson interviewed in documentary *Cover-Up at Ground Zero*. Quote from author's transcription.

38. S. Johnson interview; and Q. Nisson interview. For southern Utah tourism growth, see the following *Iron County (Cedar City UT) Record* articles: "Utah Scenic Wonders Praised in Auto Club Magazine," 20 August 1953; "Utah to Be Featured in National Business Magazine," 20 August 1953; "Utah's National Parks, Monuments, Set Visitor Records," 7 January 1954; and "Utah Tourist Council Releases New Vacation Booklet," 24 June 1954. For a history of the Shakespeare Festival, see Utah Shakespeare Festival, "About Us: History."

39. See, for example, *Iron County (Cedar City UT) Record* for 10 March 1955, 24 March 1955, 19 July 1956, 31 May 1962, and 23 August 1962. For a front-page side-by-side example, see *Iron County (Cedar City UT) Record* for 10 and 24 March 1955.

40. Author's survey of *Washington County (St. George UT) News* and *Iron County (Cedar City UT) Record* coverage, 1951–62, with assistance of Theresa Trebon.

41. Miles, "Great Fallout-Cancer Story," 3. See downwinder Mary Dickson's response, "My View: Fallout Victimization." In 2009, Miles self-published his argument in a paperback book titled *The Phantom Fallout-Induced Cancer Epidemic*.

42. Author's survey of *Washington County (St. George UT) News* and *Iron County (Cedar City UT) Record* coverage, 1951–62, with assistance of Theresa Trebon.

43. Thomas interview with author; and Thomas, phone conversation with author, 25 July 2013.

44. Scharnberg, "St. George Is Expendable," 60.

6. CRITICAL MASS

1. "Congress Should Probe," *Deseret (Salt Lake City) News*, 15 February 1979.
2. White, "Deaths High."
3. Matheson quoted in Fowler, "Scott Matheson." See also Ellis, "Cancer, Fall-out Link," clipping, folder 4, box 485, McKay Papers. Congressman Gunn McKay's press clipping file on the downwinders offers a thorough assortment of the *Deseret (Salt Lake City) News*' downwinder coverage during the late 1970s and early 1980s. McKay was a powerful political advocate for the downwinders and pushed hard for the passage of compensation legislation.
4. "Woman Fights for Safer Radiation Controls," *Deseret (Salt Lake City) News*, 10 July 1978, clipping, folder 4, box 485, McKay Papers.
5. Tsosie interview in Brugge, Benally, and Yazzie-Lewis, *Navajo People*, 161.
6. Brugge and Goble, "History of Uranium Mining," in Brugge, Benally, and Yazzie-Lewis, *Navajo People*, 36; and Eichstaedt, *If You Poison Us*, 102.
7. Francis Paul Prucha offers a comprehensive survey of the effects of the termination policy in *The Great Father*, 340–56.
8. Geneal Anderson, interview by Carole Gallagher, April 1991, Cedar City UT, in Gallagher, *American Ground Zero*, 229. For additional discussion of the impact of termination on the Southern Utah Paiute, see Holt, *Beneath Red Cliffs*, especially 61–97.
9. University of Arizona Library, "Stewart L. Udall"; and Udall interview by author.

10. Clarkin, *Federal Indian Policy*, 244. For an eloquent Native perspective on the long-term effects of the Peabody strip mining operation on Black Mesa, see interview with Wahleah Johns (Navajo) and Lilian Hill (Hopi), founders of the Black Mesa Water Coalition, by Paget-Clarke.

11. Clarkin, *Federal Indian Policy*, 153–54.

12. Udall, *Myths of August*, 184, 185.

13. This uranium boom heavily affected native and nonnative communities alike in the Four Corners area, elsewhere in the American West, and internationally. For the purposes of this work, I focus primarily on the Navajo and Laguna people, but resources on other affected communities include Amundson, *Yellowcake Towns*, SRIC (Social Issues Research Centre), online at http://www.sric.org/ (accessed 16 July 2009), and the SHAWL (Sovereignty, Health, Air, Water, Land) Society, online at http://shawlsociety .blogspot.com/ (accessed 16 July 2009).

14. Wagoner et al., "Radiation as the Cause," 181–88; and Statement of Hon. Willard W. Wirtz, Secretary of Labor, Hearings before the Subcommittee on Research, Development, and Radiation of the Joint Committee on Atomic Energy, Congress of the United States, 90th Cong., 1st sess., on Radiation Exposure of Uranium Miners, 1967, in Eichstaedt, *If You Poison Us*, appendix 3, 219.

15. Molly Ivins, "Uranium Mines Leaving Indians a Legacy of Death," *New York Times*, 20 May 1979, quoted in Eichstaedt, *If You Poison Us*, 106; and Shuey, "Widows of Red Rock," 5–7.

16. Nelkin, "Native Americans and Nuclear Power," 5.

17. "Califano Orders Search for Cancer Studies," *Deseret (Salt Lake City) News*, 8 January 1979; White, "AEC Put Lid"; Dr. Joseph Lyon, testimony, "Health Risks Associated with Atomic Tests in the 1950s," 8, Hearing of Labor, Health and Human Services, and Education Subcommittee of Senate Appropriations, 1 October 1997 (hereafter HRAAT), Seegmiller Papers; and Sternglass. "Cancer," 1102–4.

18. Gofman, *Radiation and Human Health*, 107; and Dawson and Madsen, "Uranium Mine Workers," in Johnston, *Half-Lives*, 127.

19. Lyon, testimony, HRAAT, 8; and Lyon et al., "Childhood Leukemias," 397.

20. "Congress Should Probe Utah's Fallout Problem," *Deseret (Salt Lake City) News*, 15 February 1979. See, for example: White, "Never Lost"; Ivins, "Utah Uneasy"; "Study Shows High Leukemia Toll in Area of Utah Probably Hit by Nuclear Fallout" *Wall Street Journal*, 23 February 1979; "Utah Children Born during Atomic Testing Suffered 40% Higher Leukemia Death Rates,"

Washington County (St. George UT) News, 1 March 1979; and Curry, "After A-Test Accident."

21. For citizens' stories of downwind contamination from the Three Mile Island region, see De Tredici, *People of Three Mile Island*. Also see IMDB, "*The China Syndrome*."

22. "Hatch Confirms Hearings in Salt Lake, Apologizes for 'Confusion,'" *Washington County (St. George UT) News*, 19 April 1979; and Raine and Nelson, "Public's Health Neglected."

23. Raine and Nelson, "Public's Health Neglected."

24. Raine and Nelson, "Public's Health Neglected"; and Dvorchak, "Stricken Reactor."

25. Committee members' names are from See, "Committee Organized," clipping in folder 4, box 485, McKay Papers. See also See, "Fallout Aimed, Udall Says."

26. Garcia-Mendoza, "Of Volcanoes," in Strong, *The Maternal Is Political*, 66. Bullard quoted in Cowell, "In Their Backyard," 7.

27. Bullard quoted in Cowell, "In Their Backyard," 7; and Abzug quoted by Davis, *When Smoke Ran*, 160.

28. Peterson interview in Gallagher, *American Ground Zero*, 126.

29. Brower interview in Gallagher, *American Ground Zero*, 264.

30. Phillips, interview in Gallagher, *American Ground Zero*, 303.

31. Fischer, *Citizens, Experts*, 52. The *Allen* lawsuit was a collaboration between Udall and attorney Dale Haralson of Tucson, Arizona, who had begun organizing a legal action on behalf of four Arizona downwinder families several months before Udall became involved. For a discussion of Udall and Haralson's sometimes contentious collaboration in preparing the case, see Fradkin, *Fallout*, 28–55. For additional background about the organization of the *Allen* case, see "Fallout Caseload Stuns Lawyers."

32. Plaintiffs' complaint, section 7, lines 15–26, *Irene Allen*, submitted 30 August 1979, McKay Papers, Mss 86, box 485, FD 10. See, "Committee Organized"; and See, "Fallout Caseload Stuns Lawyers." See Judge Bruce Jenkins, Memorandum Opinion, *Irene Allen*. For discussions of the case, see Fradkin, *Fallout*, 228–34; and Titus, *Bombs in the Backyard*, 119–21. Jenkins quote from Ball, *Justice Downwind*, 155.

33. For records of the hearing, see U.S. House, *Health Effects*. Quote from Christensen, "Memorandum, Findings of Fact," 23. See also Titus, *Bombs in the Backyard*, 123.

34. Beerman, *Administrative Law 2010*, 194.

35. Gottlieb and Husen, "Lung Cancer," 449–52; "Conditions Foster," *Navajo-Hopi Observer* (Flagstaff AZ), 25 March 1982; and Samet et al., "Uranium Mining," 1481–84.

36. The "cries for redress" quote can be found in the Advisory Committee on Human Radiation Experiments, *Final Report*, 577. Stewart Udall recounts the decision in *Myths of August*, 200; and Annas is quoted by Udall, *Myths of August*, 200–201.

37. Quote from Judge J. Blaine Anderson, "Overview," 13 August 1985, footnote 3, paragraph 40, Appeal of Begay et al. v. United States, available online at https://bulk.resource.org/courts.gov/c/f2/768/768.f2d.1059.84–2462.html (accessed 23 December 2012).

38. Lyon, testimony, HRAAT, 9. Dr. Billings Brown also felt the study understated exposure and disease rates. See his interview with Carole Gallagher, Holladay UT, in Gallagher, *American Ground Zero*, 260.

39. Schoemehl, "Nuclear Reactions," 24, 4.

40. Cameron, *Informal Sociology*, 13; Brown and Mikkelsen, *No Safe Place*, online excerpt of chap. 4, p. 7; and Beverly Paigen, "Controversy at Love Canal," *Hastings Center Report* 12, no. 3 (1982): 32, quoted in Brown and Mikkelsen, *No Safe Place*, online excerpt of chap. 4, pp. 7–8.

41. Pendleton deposition, 19, 20, 34–36.

42. Lyon, testimony, HRAAT, 8; and Davis, *When Smoke Ran*, xx.

43. Peter Eichstaedt discusses the reclamation crisis on the reservation lands of the Four Corners in great detail, incorporating a wealth of oral narrative from those involved in and affected by reclamation or the lack thereof. See Eichstaedt, *If You Poison Us*, 127–49. Bullard is quoted in Cowell, "In Their Backyard," 7.

44. "Citizens Call in Utah Holds Meeting in St. George," *Washington County (St. George UT) News*, 3 January 1980.

45. United States Department of Justice, "Radiation Exposure Compensation Act."

46. United States Department of Justice, "Radiation Exposure Compensation Act."

47. United States Department of Justice, "Radiation Exposure Compensation Act."

48. Gofman interview in Gallagher, *American Ground Zero*, 326. See also Parascandola, "Uncertain Science," 559–84; and C. Peterson interview.

49. Quotes from Preston Jay Truman, e-mail to author, 9 August 2005. Emphasis in original.

50. Dawson, "Navajo Uranium Workers," 394–95.

51. Dawson, "Navajo Uranium Workers," 394–95.

52. Jack interview in Brugge, Benally, and Yazzie-Lewis, *Navajo People*, 54.

53. Zang interview.

54. Fogarty, Ledesma, and Shuey, "Educating the Next Generation."

55. Dr. Richard Klausner, testimony, HRAAT, 5, 6.

56. See National Cancer Institute, "Estimated Exposures."

57. Lyon, testimony, HRAAT, 10; and Dr. Jan Beyea, testimony, HRAAT, 11, 12.

58. Timothy Conner, testimony, HRAAT, 15; and Pendleton deposition, 35–36.

59. Fischer, *Citizens, Experts*, 53.

60. See Squitieri, "Bush Pushes for New Nukes"; Editorial, "Our Worst Nightmare," *Salt Lake Tribune*, 28 September 2004; and Goodman, "U.S. Enters New Nuclear Age."

61. For sponsors, see welcome pamphlet, "Atomic Testing Museum" (Las Vegas: Atomic Testing Museum, 10 June 2005). The author obtained a copy during a tour of the museum 21 October 2005.

62. Ward, "Fireworks Deferred."

63. "Divine Victory: Downwinders 1, Federal Government 0," *Salt Lake Tribune*, 23 February 2007; and Union of Concerned Scientists, "Complex 2030."

64. Parenti, "What Nuclear Renaissance?," 11–16; Lewis, "Nuclear Option," 56–63, 92; Pasternak, "Mining Firms"; World Nuclear Association, "In Situ Leach (ISL) Mining"; and Pino interview.

65. Pasternak, "Mining Firms."

66. Bullard quoted in Cowell, "In Their Backyard," 9.

67. See Mining Watch Canada, "Declaration." Statements on its reception are from Pino, "Riding the Yellowcake Road."

68. Berry, "Pleasures of Eating," 148.

CONCLUSION

1. In 1987, Ferenc Szasz did state in writing that "all of us have become 'downwinders.'" See Szasz, "Downwind from the Bomb," 187.

2. Berry, "Work of Local Culture," 154–55.

3. Davis, *Secret History*, xii.

4. Gunderson quoted by Jamail, "Fukushima." See also Tabuchi, "Angry Parents in Japan"; Sulzberger and Wald, "Flooding Brings Worry"; Rushe, "Wildfire Reaches Los Alamos"; Dedman, "General Electric–Designed Reactors"; and McCurry, "Fukushima's Children."

5. Confino, "Collective Memory," 1401–2; and White, *Remembering Ahanagran*, 152.

BIBLIOGRAPHY

UNPUBLISHED SOURCES

Benally, Timothy. Interview by author. Shiprock NM, 11 January 2008.

Bulloch, McRae. Interview by author. Cedar City UT, 20 October 2005.

Dickson, Mary. Interview by author. Salt Lake City UT, 26 October 2005.

Harrison, Phil. Interview by author. Shiprock NM, 11 January 2008.

Johnson, Loa. Interview by author. St. George UT, 19 October 2005.

Johnson, Sheldon, and LaVerna Johnson. Interview by author. St. George UT, 19 October 2005.

McKay, Gunn, Papers. Special Collections. Utah State University, Logan.

Nisson, Quentin, and Gwendolyn Nisson. Interview by author. Washington City UT, 21 October 2005.

Peterson, Claudia. Interview by author. St. George UT, 20 October 2005.

Peterson, Willard "Bill." Interview by author. Logan UT, 19 January 2006.

Pino, Manuel "Manny." Interview by author. Scottsdale AZ, 9 January 2008.

Seegmiller, Janet. Interview by author. Cedar City UT, 18 October 2005.

Sidney, Ivan. Interview by author. Kykotsmovi Village AZ, 13 January 2008.

Thomas, Michelle. Essay, "Meditations on My Mother," 2005.

———. Interview by author. St. George UT, 20 October 2005.

Thorely, Roe. Cedar City UT, 17 March 1981, transcript in Oral History Collection, Special Collections, Southern Utah University, Cedar City.

Timothy, Dave. Interview by author. Toquerville UT, 18 October 2005.

Truman, Jay Preston. Interview by author. Malad ID, 2 September 2005.

Udall, Stewart. Interview by author. Santa Fe NM, 25 May 2003.

Verdoia, Ken. Interview by author. Salt Lake City UT, 12 July 2005.

Zang, Douglas. Interview by author. Shiprock NM, 12 January 2008.

PUBLISHED SOURCES

Aarne, Antti. *The Types of the Folktale: A Classification and Bibliography.* Helsinki: Academia Scientiarum Fennica, 1961.

Advisory Committee on Human Radiation Experiments. *Final Report.* October 1995. https://archive.org/details/advisorycommittee0ounit. Accessed 3 April 2014.

———. *Interim Report of October 21, 1994.* http://www.gwu.edu/~nsarchiv/radiation/. Accessed 16 June 2011.

Alexander, Thomas G. "Irrigating the Mormon Heartland: The Operation of the Irrigation Companies in the Wasatch Oasis Communities, 1847–1880." *Agricultural History* 76 (Spring 2002): 172–87.

———. "Stewardship and Enterprise: The LDS Church and the Wasatch Oasis Environment, 1847–1930." *Western Historical Quarterly* 25 (Autumn 1994): 340–64.

Allen, Barbara. "The Personal Point of View in Orally Communicated History." *Western Folklore* 38 (April 1979): 110–18.

Alperovitz, Gar. *The Decision to Use the Atomic Bomb.* New York: Vintage Books, 1995.

———. "Enola Gay: Was Using the Bomb Necessary?" *Miami Herald,* 14 December 2003. http://www.commondreams.org/scriptfiles/views03/1214–06.htm. Accessed 30 March 2008.

American Society of Mechanical Engineers International and Thermo King Corporation. *Thermo King Model C Transport Refrigeration Unit: An International Historic Mechanical Engineering Landmark.* Pamphlet. New York: American Society of Mechanical Engineers International, October 1996.

Amundson, Michael A. *Yellowcake Towns: Uranium Mining Communities in the American West.* Boulder: University of Colorado Press, 2002.

Anaya, S. James. "International Law and U.S. Trust Responsibility toward Native Americans." In *Native Voices: American Indian Identity and Resistance,* edited by Richard A. Grounds, George E. Tinker, and David E. Wilkins, 155–85. Lawrence: University Press of Kansas, 2003.

Atomic Archive. "Nuclear Test Sites." http://www.atomicarchive.com/Almanac /Testing.shtml. Accessed 2 October 2008.

Ball, Howard. *Cancer Factories: America's Tragic Quest for Uranium Self-Sufficiency.* Westport CT: Greenwood Press, 1993.

———. *Justice Downwind: America's Atomic Testing Program in the 1950s.* New York: Oxford University Press, 1986.

Basso, Keith. *Wisdom Sits in Places: Landscape and Language among the Western Apache.* Albuquerque: University of New Mexico Press, 1996.

Bauman, Richard. *Verbal Art as Performance.* Prospect Heights IL: Waveland Press, 1977.

Beck, Colleen, M. Nieves Zedeño, and Robert Furlow. "Time, Trust, and the Measure of Success: The Nevada Test Site Cultural Resources Program." *Society for American Archaeology Bulletin* 15 (May 1997). http://www.saa .org/Portals/0/SSA/publications/SAAbulletin/15–3/SAA23.html. Accessed 23 February 2009.

Beerman, Jack M. *Administrative Law 2010.* New York: Aspen Publishers, 2010.

Begay et al. v. United States, No. 84-2462, United States District Court for the District of Arizona, Phoenix (9th Cir., August 1985).

Benally, Timothy. "Navajo Uranium Miners Fight for Compensation." *In Motion Magazine,* 20 September 1999. http://www.inmotionmagazine.com/miners .html. Accessed 18 October 2008.

Bernstein, Alison R. *American Indians and World War II: Toward a New Era in Indian Affairs.* Norman: University of Oklahoma Press, 1999.

Berry, Wendell. "The Pleasures of Eating." In *What Are People For?*, 145–52. London: Rider Books, 1991.

———. "The Work of Local Culture." In *What Are People For?*, 153–69. London: Rider Books, 1991.

Bordewich, Fergus M. *Killing the White Man's Indian: Reinventing Native Americans at the End of the Twentieth Century.* New York: Doubleday, 1996.

Boyer, Paul. *By the Bomb's Early Light: American Thought and Culture at the Dawn of the Atomic Age.* New York: Pantheon, 1985.

Brown, Phil, and Edwin J. Mikkelsen. *No Safe Place: Toxic Waste, Leukemia, and Community Action.* Berkeley: University of California Press, 1997.

Brugge, Doug, Timothy Benally, and Phil Harrison. *Memories Come to Us in the Rain and the Wind: Oral Histories and Photographs of Navajo Uranium Miners and Their Families.* Boston: Tufts, 1997.

Brugge, Doug, Timothy Benally, and Esther Yazzie-Lewis, eds. *The Navajo People and Uranium Mining.* Albuquerque: University of New Mexico Press, 2006.

Brugge, Doug, Jamie L. deLemos, and Cat Bui. "The Sequoyah Corporation Fuels Release and the Church Rock Spill: Unpublicized Nuclear Releases in American Indian Communities." *American Journal of Public Health*, 97, no. 9 (September 2000): 1595–1600. http://www.ncbi.nlm.nih.gov/pmc /articles/PMC1963288/. Accessed 5 April 2014.

Bulloch et al. v. United States, No. C-19-55, United States District Court for the District of Utah, Central Division, Salt Lake City, 16 March 1955.

Burke, Peter. "History as Social Memory." In *Memory: History, Culture and the Mind*, 97–113, edited by Thomas Butler. New York: Basil Blackwell, 1989.

Cameron, William Bruce. *Informal Sociology: A Casual Introduction to Sociological Thinking*. New York: Random House, 1963.

Carson, Rachel. *Silent Spring*. Boston: Houghton Mifflin, 1962.

Choudhary, Trib. "The Navajo Nation: 2000–2001 Comprehensive Economic Development Strategy." http://www.navajobusiness.com/pdf/CEDS /CEDS%202002-03.pdf. Accessed 10 July 2008.

Clarfield, Gerard H., and William M. Wiecek. *Nuclear America: Military and Civilian Nuclear Power in the United States, 1940–1980*. New York: Harper & Row, 1984.

Clark, Blue. *Lone Wolf v. Hitchcock: Treaty Rights and Indian Law at the End of the Nineteenth Century*. Lincoln: University of Nebraska Press, 1994.

Clarkin, Thomas. *Federal Indian Policy in the Kennedy and Johnson Administrations, 1961–1969*. Albuquerque: University of New Mexico Press, 2001.

Colorado Plateau–Land Use History of North America. "Cattle and Sheep Grazing: History." http://www.cpluhna.nau.edu/Change/grazing.htm. Accessed 20 March 2006.

———. "Modern Climactic Conditions on the Colorado Plateau." http://cpluhna .nau.edu/Change/modern_climatic_conditions.htm, Accessed 20 April 2009.

Committee to Assess Health Risks from Exposure to Low Levels of Ionizing Radiation, National Research Council. *Health Risks from Exposure to Low Levels of Ionizing Radiation: BEIR VII Phase 2*. Washington DC: National Academies Press, 2006.

Confino, Alon. "Collective Memory and Cultural History: Problems of Method." *American Historical Review* 102 (December 1997): 1386–403.

Consumers' Union. "The Milk We Drink." *Consumer Reports*, March 1959, 102–11.

Cordell, L. *Archaeology of the Southwest*. 2nd ed. San Diego: Academic Press, 1997.

The Cover-Up at Ground Zero. Produced and directed by Elena Mannes, Turning Point series, ABC. 1994. Videocassette.

Cowell, Rebekah. "In Their Backyard: Robert D. Bullard on the Politics of Where We Put Our Trash." *The Sun* 437 (May 2012): 5–10.

Curry, Bill. "After A-Test Accident, Safety Didn't Come First." *Washington Post*, 2 March 1979.

D'Antonio, Michael. *Atomic Harvest: Hanford and the Lethal Toll of America's Nuclear Arsenal.* New York: Crown, 1993.

David L. Timothy et al. v. United States, No. C-80-0445 A, United States District Court for the District of Utah, Central Division, Salt Lake City.

Davidson, Lee, and Joe Bauman. "Toxic Utah: A Land Littered with Poisons." *Deseret (Salt Lake City) News*, 28 February 2001. http://www.deseretnews .com/article/1,5143,250010322,00.html?pg=1. Accessed 10 October 2008.

Davis, Devra. *The Secret History of the War on Cancer.* New York: Basic Books, 2007.

———. *When Smoke Ran Like Water: Tales of Environmental Deception and the Battle against Pollution.* New York: Basic Books, 2002.

Dawson, Susan E. "Navajo Uranium Workers and the Effects of Occupational Illness: A Case Study." *Human Organization* 51 (Winter 1992): 389–97.

Dawson, Susan E., Perry H. Charley, and Phillip Harrison Jr. "Advocacy and Social Action among Navajo People: Uranium Workers and Their Families, 1988–1995." In Brugge, Benally, and Yazzie-Lewis, *Navajo People*, 69.

Dawson, S. E., and G. E. Madsen. "American Indian Uranium Millworkers: The Perceived Effects of Chronic Occupational Exposure." *Journal of Health and Social Policy* 7, no. 2 (1995): 19–31.

———. "Uranium Mine Workers, Atomic Downwinders, and the Radiation Exposure Compensation Act (RECA): The Nuclear Legacy." In Johnston, *Half-Lives*, 117–43.

Dawson, S. E., G. E. Madsen, and B. R. Spykerman. "Public Health Issues Concerning American Indian and Non-Indian Uranium Millworkers." *Journal of Health and Social Policy* 8, no. 3 (1997): 41–56.

Dedman, Bill. "General Electric–Designed Reactors Have 23 Sisters in U.S." *Open Channel Blog*, 13 March 2011. http://openchannel.msnbc.msn.com /_news/2011/03/13/6256121-general-electric-designed-reactors-in -fukushima-have-23-sisters-in-us. Accessed 18 July 2011.

Deloria, Phillip J. *Indians in Unexpected Places.* Lawrence: University of Kansas Press, 2004.

Del Tredici, Robert. *The People of Three Mile Island.* San Francisco: Sierra Club Books, 1980.

Dibblin, Jane. *A Day of Two Suns: U.S. Nuclear Testing and the Pacific Islanders.* New York: New Amsterdam Press, 1990.

Dickson, Mary. "My View: Fallout Victimization Absolutely Not Exaggerated." *Deseret (Salt Lake City) News,* 15 November 2005. http://findarticles.com/p /articles/mi_qn4188/is_/ai_n15836498. Accessed 18 October 2008.

Didion, Joan. *The White Album.* New York: Simon & Schuster, 1979. Reprint, New York: Farrar, Straus and Giroux, 1990.

Dillon, Lucinda. "Toxic Utah: Ghosts in the Wind." *Deseret (Salt Lake City) News,* 15 February 2001. http://www.deseretnews.com/article/250011123 /Toxic-Utah-Ghosts-in-the-wind.html. Accessed 15 July 2013.

Dreicer, Mona, André Bouville, and Bruce W. Wachholz. "Pasture Practice, Milk Distribution, and Consumption in the Continental U.S. in the 1950s." *Health Physics* 59, no. 5 (1990): 627–36.

Dvorchak, Bob. "Stricken Reactor Cools below Boiling Point, Months from Cold Shutdown Condition." *Salt Lake Tribune,* 21 April 1979.

Dyson, Freeman. "Comments on Sternglass's Thesis." *Bulletin of the Atomic Scientists,* June 1969, 27.

Eichstaedt, Peter. *If You Poison Us: Uranium and Native Americans.* Santa Fe: Red Crane Books, 1994.

Ellis, Dexter. "Cancer, Fallout Link Concerns Matheson." *Desert (Salt Lake City) News,* 29 August 1977.

Embassy of the Republic of the Marshall Islands. "Nuclear Issues." http://www .rmiembassyus.org/Nuclear%20Issues.htm. Accessed 17 October 2008.

Esson, Dylan Jim. "Did 'Dirty Harry' Kill John Wayne? Media Sensationalism and the Filming of *The Conqueror* in the Wake of Atomic Testing." *Utah Historical Quarterly* 71 (Summer 2003): 250–65.

Evans, Robley. "Medical Uses of Atomic Energy." *Atlantic Monthly,* January 1946, 63–68.

Fadiman, Anne. "The Downwind People." *Life Magazine,* June 1980, 32–40.

Federation of American Scientists. "Plot of Worldwide Nuclear Tests." 2008. http://www.fas.org/programs/ssp/nukes/testing/plotworldtests.html. Accessed 15 June 2011.

Ficken, Robert. "Grand Coulee and Hanford: The Atomic Bomb and the Development of the Columbia River." In Hevly and Findlay, *The Atomic West,* 21–38.

Fischer, Frank. *Citizens, Experts, and the Environment: The Politics of Local Knowledge.* Durham NC: Duke University Press, 2000.

Fleischner, Tom. "Ecological Costs of Livestock Grazing in Western North America." *Conservation Biology* 8 (1994): 629–44.

Fogarty, John, Sylvia Ledesma, and Chris Shuey. "Educating the Next Generation of Doctors: Rx for Undoing Racism." *Voices from the Earth* 5 (Spring 2004). http://www.sric.org/voices/2004/v5n1/undoing_racism.html. Accessed 15 October 2008.

Fowler, Glenn. "Scott Matheson, 61, Ex-Governor and Leading Democrat in Utah." *New York Times*, 8 October 1990, D10.

Fradkin, Phillip. *Fallout: An American Nuclear Tragedy.* Tucson: University of Arizona Press, 1989.

Francaviglia, Richard. *Believing in Place: A Spiritual Geography of the Great Basin.* Reno: University of Nevada Press, 2003.

Frisch, Michael. *A Shared Authority: Essays on the Craft and Meaning of Oral and Public History.* Albany: State University of New York Press, 1990.

Fuller, John. *The Day We Bombed Utah: America's Most Lethal Secret.* New York: New American Library, 1984.

Gallagher, Carole. *American Ground Zero: The Secret Nuclear War.* Cambridge MA: MIT Press, 1993.

Garcia-Mendoza, Violeta. "Of Volcanoes and Ruins and Gardens." In *The Maternal Is Political: Women Writers at the Intersection of Motherhood and Social Change,* edited by Shari MacDonald Strong, 59–66. Berkeley: Seal Press, 2008.

Geertz, Clifford. *After the Fact: Two Countries, Four Decades, One Anthropologist.* Cambridge MA: Harvard University Press, 1995.

———. *Local Knowledge: Further Essays in Interpretive Anthropology.* New York: Basic Books, 1983.

Glass, Matthew. "Air Force, Western Shoshone, and Mormon Rhetoric of Place and the MX Conflict." In Hevly and Findlay, *The Atomic West,* 255–75.

Glazer, Myron P., and Penina M. Glazer. "National Security and the Citizen's Crusade for a Safe Environment in the United States, Israel, and Czechoslovakia." *Massachusetts Review* 37, no. 2 (1996): 271–304.

Gofman, John W. *Radiation and Human Health: A Comprehensive Examination of the Evidence Relating Low-Level Radiation to Cancer and Other Diseases.* San Francisco: Sierra Club Books, 1981.

Goodman, Amy. "U.S. Enters New Nuclear Age as Bush Seeks Funds for New Generation of Nukes." *Democracy Now,* 2 March 2006. Transcript, http://www.democracynow.org/article.pl?sid=06/03/02/148233& mode=thread &tid=25. Accessed 2 April 2006.

Goodman, David, and Amy Goodman. "Hiroshima Cover-up: Stripping the War Department's Timesman of His Pulitzer." *Democracy Now,* 5 August 2005.

http://www.democracynow.org/2005/8/5/hiroshima_cover_up_stripping _the_war. Accessed 21 August 2012.

Gottleib, Leon S., and LaVerne A. Husen. "Lung Cancer among Navajo Uranium Miners." *CHEST* 81 (April 1982): 449–52.

Goulding, Harry. "Navajos Hunt Big Game . . . Uranium." *Popular Mechanics*, June 1950, 89–92, 236–40.

Griggs, Brandon. "'Exposed's' Requiem for the Dead Continues to Grow." *Salt Lake Tribute*, 3 November 2007. http://www.sltrib.com/portlet/article /html/fragments/print_article.jsp?articleId=7362284&siteId=297. Accessed 28 August 2008.

Gunderson, Gordon W. "The National School Lunch Program: Background and Development." http://www.fns.usda.gov/cnd/lunch/AboutLunch/Program History_5.htm. Accessed 6 October 2008.

Hacker, Barton C. *Elements of Controversy: The Atomic Energy Commission and Radiation Safety in Nuclear Weapons Testing, 1947–1974.* Berkeley: University of California Press, 1994.

———. "'Hotter Than a $2 Pistol': Fallout, Sheep, and the Atomic Energy Commission, 1953–1986." In Hevly and Findlay, *The Atomic West*, 157–75.

Hammond, E. C., and L. Garfinkel. "Changes in Cigarette Smoking, 1959–1965." *American Journal of Public Health* 58 (January 1968): 30–45.

Harney, Corbin. *The Way It Is: One Water . . . One Air . . . One Mother Earth.* Nevada City CA: Blue Dolphin Publishing, 1995.

Havlice, Patricia Pate. *Oral History: A Reference Guide and Annotated Bibliography.* Jefferson NC: McFarland, 1985.

Helms, Kathy. "Red Valley Keeps an Eye on RECA." *Gallup Independent*, 27 August 2008. http://www.gallupindependent.com/2008/08august/082708redvaley .html. Accessed 1 September 2008.

Hennessey, Kathleen. "Plans for a Massive Blast in Nevada Draw Fire." *Associated Press*, 31 March 2006. http://www.commondreams.org/headlines06/0331-04 .htm. Accessed 31 March 2006.

Hevly, Bruce, and John M. Findlay, eds. *The Atomic West.* Seattle: Center for the Study of the Pacific Northwest and University of Washington Press, 1998.

———. "The Atomic West: Region and Nation, 1942–1992." In Hevly and Findlay, *The Atomic West*, 3–18.

Holl, Jack M., and Terrence R. Fehner. Review of *Bombs in the Backyard: Atomic Testing and American Politics* by A. Constandina Titus. *Journal of American History* 74, no. 3 (December 1987): 1093.

Holt, Ronald. *Beneath These Red Cliffs: An Ethnohistory of the Utah Paiutes.* Logan: Utah State University Press, 2006.

Hopi Cultural Preservation Office. "Hopi Agriculture." http://www.nau .edu/~hcpop/culture/agric.htm. Accessed 18 October 2008.

Hunner, Jon. *Inventing Los Alamos: The Growth of an Atomic Community.* Norman: University of Oklahoma Press, 2004.

International Oral History Association. "Homepage." http://www.ioha.fgv.br/ioha /english/index.html. Accessed 29 March 2006.

Internet Movie Database (IMDB). "*The China Syndrome.*" http://www.imdb.com /title/tt0078966/. Accessed 10 July 2009.

———. "*Godzilla, King of the Monsters.*" http://www.imdb.com/title/tt0197521/. Accessed 15 September 2008.

Irene Allen et al. v. United States, No. C79-0515-J, United States District Court for the District of Utah, Central Division, Salt Lake City. 588 F Supp. 247 (1984).

Iverson, Peter. *Diné: A History of the Navajos.* Albuquerque: University of New Mexico Press, 2002.

Ives, Edward D. *The Tape-Recorded Interview: A Manual for Fieldworkers in Folklore and Oral History.* 2nd ed. Knoxville: University of Tennessee Press, 1995.

Ivins, Molly. "Utah Uneasy over Leukemia-Atomic Tests Study; Government Facing Lawsuits." *New York Times*, 23 February 1979.

Jackson, Bruce, and Edward Ives, eds. *The World Observed: Reflections on the Fieldwork Process.* Urbana: University of Illinois Press, 1996.

Jamail, Dahr. "Fukushima: It's Much Worse Than You Think." *Al Jazeera English*, 16 June 2011. http://english.aljazeera.net/indepth/features/2011 /06/201161664828302638.html. Accessed 19 July 2011.

Japan Peace Museum. *The Nuclear Century: Voices of the Hibakusha of the World.* Tokyo: Japan Peace Museum, 1997.

Jensen, Derrick. *The Culture of Make Believe.* New York: Context Books, 2002.

Johnston, Barbara Rose. "Half-Lives, Half-Truths, and Other Radioactive Legacies of the Cold War." In Johnston, *Half-Lives*, 1–23.

———, ed. *Half-Lives & Half-Truths: Confronting the Radioactive Legacies of the Cold War.* Santa Fe: School for Advanced Research Press, 2007.

———. "'More Like Us Than Mice': Radiation Experiments with Indigenous Peoples." In Johnston, *Half-Lives*, 25–54.

Johnston, Barbara Rose, and Susan Dawson. "Resource Use and Abuse on Native American Land: Uranium Mining in the American Southwest." In *Who Pays the Price? The Sociocultural Context of Environmental Crisis*, edited by Barbara Rose Johnston, 142–53. Washington DC: Island Press, 1994.

Jones, Suzi. "Regionalization: A Rhetorical Strategy." *Journal of Folklore* 13, no. 1 (1976): 105–20.

Kansteiner, Wulf. "Finding Meaning in Memory: A Methodological Critique of Collective Memory Studies." *History and Theory* 41 (May 2002): 179–97.

Knox, Richard. "How Risky Is Infant Formula Made with Tokyo Tapwater?" *Shots: NPR.org's Health Blog*. 23 March 2011. http://www.npr.org/blogs/health /2011/03/24/134804037/how-risky-is-infant-formula-made-with-tokyo-tap -water. Accessed 17 June 2011.

Kramer, Becky. "Spokane Tribe Members Worked Gladly in Uranium Mines." *Spokesman-Review* (Spokane WA), 5 June 2011. http://www.spokesman.com /stories/2011/jun/05/i-watch-them-die-young-and-old/. Accessed 15 September 2012.

Kricher, John C. *Ecology of Western Forests*. Boston: Houghton Mifflin, 1993.

Krohn, Kaarle. *Folklore Methodology*. Translated by Roger Welsch. Cambridge MA: Harvard University Press, 1926. Reprint, Austin: University of Texas Press, 1971.

Kuletz, Valerie. *The Tainted Desert: Environmental and Social Ruin in the American West*. New York: Routledge, 1998.

Lahmeyer, Jan. "Historical Demographical Data of the Urban Centers: Utah." http://www.populstat.info/. Accessed 13 March 2006.

Lewis, Judith. "The Nuclear Option." *Mother Jones*, May/June 2008, 56–63, 92.

Life. "The 36-Hour War," 19 November 1945, 27–35.

Limerick, Patricia Nelson. *Legacy of Conquest: The Unbroken Past of the American West*. New York: W. W. Norton, 1987.

Lindee, M. Susan. *Suffering Made Real: American Science and the Survivors at Hiroshima*. Chicago: University of Chicago Press, 1994.

Liptak, Adam. "Inmate Count in U.S. Dwarfs Other Nations'." *New York Times*, 23 April 2008. http://www.nytimes.com/2008/04/23/us/23prison.html? _r=1&ref=opinion. Accessed 9 April 2009.

Lloyd, R. D., R. C. Pendleton, C. W. Mays, and D. O. Clark. "A Comparison of the Cesium-137 Content of Milk and People from 19 Dairy Farms in Utah." *Radiological Health Data and Reports* 10 (October 1969): 427–33.

Lyman, Nora. Observation. *Washington County (St. George) News*, 11 April 1955.

Lyon, Joseph L., M. R. Klauber, J. W. Gardner, and K. S. Udall. "Childhood Leukemias Associated with Fallout from Nuclear Testing." *New England Journal of Medicine* 300 (February 1979): 397–402.

Maiser, Jen, Jessica Prentice, Sage Van Wing, and DeDe Sampson. "Locavores." 13 July 2008. http://www.locavores.com/. Accessed 1 October 2008.

Makhijani, Arjun, Howard Hu, and Katherine Yih, eds. *Nuclear Wastelands: A Global Guide to Nuclear Weapons Production and Its Health and Environmental Effects*. Cambridge MA: MIT Press, 1995.

Makhijani, Arjun, and Scott Saleska. "The Production of Nuclear Weapons and Environmental Hazards." In Makhijani, Hu, and Yih, *Nuclear Wastelands*, 22–63.

Markstrom, Carol A., and Perry H. Charley. "Psychological Effects of Technological/Human-Caused Environmental Disasters: Examination of the Navajo and Uranium." *American Indian and Alaska Native Mental Health Research* 11 (2003): 19–45. Reprinted in Brugge, Benally, and Yazzie-Lewis, *Navajo People*, 89–115.

May, Elaine Tyler. *Homeward Bound: American Families in the Cold War Era*. New York: Basic Books, 1988.

McCurry, Justin. "Fukushima's Children at Centre of Debate over Rates of Thyroid Cancer." *The Guardian*, 9 March 2014. http://www.theguardian.com/world/2014/mar/09/fukushima-children-debate-thyroid-cancer-japan-disaster-nuclear-radiation. Accessed 2 April 2014.

McEnaney, Laura. *Civil Defense Begins at Home: Militarization Meets Everyday Life in the Fifties*. Princeton NJ: Princeton University Press, 2000.

McNichol, Dan. *The Roads that Built America: The Incredible Story of the U.S. Interstate System*. New York: Sterling, 2006.

Micale, Mark S. "Hysteria and Its Historiography: A Review of Past and Present Writings." *History of Science* 27 (1989): 223–61.

Miles, Daniel W. "The Great Fallout-Cancer Story of 1978 and Its Aftermath." *Forum on Physics and Society* 35. http://www.aps.org/units/fps/newsletters/2005/october/articles.cfm#miles. Accessed 29 March 2006.

———. *The Phantom Fallout-Induced Cancer Epidemic in Southwestern Utah: Downwinders Deluded and Waiting to Die*. Charleston SC: BookSurge, 2009.

Miller, Herman P. "Changes in Income Distribution in the United States." *Journal of the American Statistical Association* 46 (December 1951): 438–41.

Miller, Richard. *Under the Cloud: The Decades of Nuclear Testing*. New York: Free Press, 1986.

———. *The U.S. Atlas of Nuclear Fallout, 1951–1970*. Vol. 1, abr. ed. The Woodlands TX: Regis Books, 2000.

Mining Watch Canada. "Indigenous World Uranium Summit Declaration." 4 December 2006. http://www.miningwatch.ca/declaration-indigenous-world-uranium-summit. Accessed 22 September 2012.

Moorhead, Joanna. "Milking It." *The Guardian*, 15 May 2007. http://www.guardian .co.uk/business/2007/may/15/medicineandhealth.lifeandhealth. Accessed 17 June 2011.

Mozes, Alan. "Fallout from Atomic Bombs Still Causing Health Problems." *Washington Post*, 14 March 2008. http://www.washingtonpost.com/wp-dyn /content/article/2008/03/14/ ar2008031402230_pf.html. Accessed 15 March 2008.

Murphy, Richard. "The Isolated, Emotional, and Incomplete Storyteller: Message Board Narratives of Combat Veterans Living with PTSD." Paper presented at the annual meeting of the NCA 94th Annual Convention, 23 November 2008, San Diego CA. http://www.allacademic.com/meta/p255249_index .html. Accessed 4 July 2009.

Nabakov, Peter, ed. *Native American Testimony: A Chronicle of Indian-White Relations from Prophecy to the Present*. 1978. Revised, New York: Penguin Books, 1999.

National Cancer Institute. "Estimated Exposures and Thyroid Doses Received by the American People from Iodine-131 in Fallout Following Nevada Atmospheric Nuclear Bomb Tests." National Cancer Institute, 1 October 1997. http://www.cancer.gov/cancertopics/causes/i131/nci-reports. Accessed 15 August 2012.

Nelkin, Dorothy. "Native Americans and Nuclear Power." *Program on Science, Technology, & Human Values* 6 (Spring 1981): 2–13.

North Country Healthcare. "Radiation Exposure Screening & Education Program." http://www.northcountryhealthcare.org/what/community-services/radiation -exposure.php. Accessed 25 April 2011.

Nuclear Weapon Archive. "Gallery of U.S. Nuclear Tests," 6 August 2001. http://nuclearweaponarchive.org/Usa/Tests/index.html. Accessed 3 September 2008.

———. "Operation Upshot-Knothole: Harry," 19 June 2002. http://www.nuclear weaponarchive.org/Usa/Tests/Upshotk.html. Accessed 11 September 2008.

O'Neill, Colleen. *Working the Navajo Way: Labor and Culture in the Twentieth Century*. Lawrence: University Press of Kansas, 2005.

O'Neill, Dan. *The Firecracker Boys: H-Bombs, Inupiat Eskimos, and the Roots of the Environmental Movement*. New York: Basic Books, 1994.

Oral History Association. "Homepage." http://www.oralhistory.org/. Accessed 30 March 2006.

Oxford University Press. "Oxford Word of the Year: Locavore." OUPblog, 12 November 2007. http://blog.oup.com/2007/11/locavore/. Accessed 1 October 2008.

Paget-Clarke, Nic. Interview with Wahleah Johns and Lilian Hill. 13 June 2004. http://www.inmotionmagazine.com/global/wj_lh_int.html. Accessed 4 March 2007.

Parascandola, Mark. "Uncertain Science and a Failure of Trust: The NIH Radio-epidemiologic Tables and Compensation for Radiation-Induced Cancer." *Isis* 93 (2002): 559–84.

Parenti, Christian. "What Nuclear Renaissance?" *The Nation* (12 May 2008): 11–16.

Pasternak, Judy. "Mining Firms Again Eyeing Navajo Land." *Los Angeles Times*, 22 November 2006. http://www.latimes.com/news/local/la-na-navajo 22nov22,1,7362075,full.story. Accessed 14 October 2008.

———. "Oases in Navajo Desert Contained 'a Witch's Brew.'" *Los Angeles Times*, 20 November 2006. http://www.latimes.com/news/nationworld/nation/la -na-navajo20nov20,0,6106722.story. Accessed 21 November 2006.

———. "A Peril that Dwelt among the Navajos." *Los Angeles Times*, 19 November 2006. http://www.latimes.com/news/nationworld/nation/la-na-navajo 19nov19,0,1645689.story. Accessed 18 November 2008.

———. *Yellow Dirt: An American Story of a Poisoned Land and a People Betrayed.* New York: Free Press, 2010.

Pendleton, R. C., R. D. Lloyd, D. O. Clark, C. W. Mays, and G. B. Goates. "Cesium-137 in Humans: A Relationship to Milk Cs^{137} Content." *Health Physics* 24 (January 1973): 23–36.

Pendleton, R .C., C. W. Mays, R. D. Lloyd, and A. L. Brooks. "Differential Accumulation of I^{131} from Local Fallout in People and Milk." *Health Physics* 9 (December 1963): 1253–62.

Pino, Manuel. "Riding the Yellowcake Road." *Voices from the Earth* 8 (Summer 2007). http://www.sric.org/voices/2007/v8n2/index.php. Accessed 18 October 2008.

Pollan, Michael. *The Omnivore's Dilemma: A Natural History of Four Meals.* New York: Penguin, 2007.

Pope, Daniel. "Antinuclear Activism in the Pacific Northwest: WPPSS and Its Enemies." In Hevly and Findlay, *The Atomic West*, 236–54.

Propp, V. *Morphology of the Folktale.* 2nd ed. Edited by Louis A. Wagner, translated by Laurence Scott. 1968. Reprint, Austin: University of Texas Press, 1996.

Prucha, Francis Paul. *The Great Father: The United States Government and the American Indians.* Lincoln: University of Nebraska Press, 1984, 1986.

Quinton, Michael E. "The Federal Tort Claims Act: An Overview." http://www .quintonpetix.com/fedtorac.htm. Accessed 18 July 2009.

Raine, George, and Angelyn Nelson. "Public's Health Neglected in N-Test, Hearing Told." *Salt Lake Tribune*, 20 April 1979.

Rhodes, Richard. *The Making of the Atomic Bomb*. New York: Simon & Schuster, 1986.

Ringolz, Raye C. *Uranium Frenzy: Saga of the Nuclear West*. Logan: Utah State University Press, 2002.

Rizvi, Haider. "Native Americans Want 'Bunker Buster' Test Stopped." *Common Dreams*, 11 April 2006. http://www.commondreams.org/headlines06/0411-07 .htm. Accessed 11 April 2006.

Robinson, Paul. "Uranium Mill Tailings Remediation Performed by the U.S. DOE: An Overview." Southwest Research and Information Center, 18 May 2004. http://www.sric.org/uranium/rirf.php. Accessed 25 April 2011.

———. "Uranium Production and Its Effects on Navajo Communities along the Rio Puerco in Western New Mexico." In *Race and the Incidence of Environmental Hazards: A Time for Discourse*, edited by Bunyan Bryant and Paul Mohai, 153–63. Boulder: Westview Press, 1992.

Rollin, Bernard E. *Farm Animal Welfare: Social, Bioethical, and Research Issues*. Ames: Iowa State University Press, 1995.

Rondon, Anna. Testimony. World Uranium Hearings, Salzburg, Austria, 16 September 1992. http://www.ratical.org/radiation/WorldUraniumHearing /AnnaRondon.html. Accessed 17 March 2008.

Rosen, Ruth. "Flashback: New Report Proves Point Made by Mother's March in '61." *Fort Lauderdale Sun Sentinel*, 28 August 1997.

Rosenberg, Gerald. "The Sorrow and the Pity: Kent State, Political Dissent, and the Misguided Worship of the First Amendment." In *The Boundaries of Freedom of Expression and Order in American Democracy*, edited by Thomas R. Hensley, 17–37. Kent OH: Kent State University Press, 2001.

Rushe, Dominic. "Wildfire Reaches Los Alamos Nuclear Facility." *The Guardian*, 27 June 2011. http://www.guardian.co.uk/world/2011/jun/28/fire-los-alamos -nuclear-facility-new-mexico. Accessed 14 July 2011.

Saffer, Thomas H., and Orville E. Kelly. *Countdown Zero: GI Victims of U.S. Atomic Testing*. Harmondsworth UK: Penguin Books, 1983.

Samet, Jonathan M., Daniel M. Kutvirt, Richard J. Waxweiler, and Charles R. Key. "Uranium Mining and Lung Cancer in Navajo Men." *New England Journal of Medicine* 310 (7 June 1984): 1481–84.

Scharnberg, Ken. "'St. George Is Expendable.'" *The American Legion*, October 1995, 20–27, 60, 62.

Schoemehl, Frederick Anthony. "Nuclear Reactions: National Security Policy, Culture, and Environment in the Nevada Test Site Region, 1950–1958." PhD diss., University of California, Irvine, 2004.

Schrager, Sam. "The Stories Communities Tell." *Oregon Historical Quarterly* 97 (Summer 1996): 212–29.

See, Dorothy. "Committee Organized to Aid Fallout Victims." *Deseret (Salt Lake City) News*, 23 October 1978.

———. "Fallout Aimed, Udall Says." *Desert (Salt Lake City) News*, 3 October 1978.

———. "Fallout Caseload Stuns Lawyers." *Deseret (Salt Lake City) News*, 26 October 1978.

Seegmiller, Janet Burton. "The History of Iron County: Nuclear Testing and the Downwinders." Utah History to Go. http://historytogo.utah.gov/utah _chapters/utah_today/nucleartestingandthedownwinders.html. Accessed 13 July 2009.

Seidel, Robert. "Books on the Bomb." *Isis* 81 (1990): 519–37.

Shannon, M., and J. W. Graef. "Lead Intoxication from Lead-Contaminated Water Used to Reconstitute Infant Formula." *Clinical Pediatrics*, August 1989, 380–82.

Sharp, Patrick B. *Savage Perils: Racial Frontiers and Nuclear Apocalypse in American Culture*. Norman: University of Oklahoma Press, 2007.

Sharrock, Justine. "Spinning the Atom: 'Environmentalists' Are Key to the Industry's PR Push." *Mother Jones*, May/June 2008, 58–59.

Shuey, Chris. "The Widows of Red Rock." *Scottsdale (AZ) Progress, Saturday Magazine*, 9 June 1979, 5–7.

Shuey, Chris, and Melinda Ronca-Battista. *Report of the Church Rock Uranium Monitoring Project (CRUMP), 2003–2007*, Prepared 2007. http://www.sric .org/uranium/docs/CRUMPReportSummary.pdf. Accessed 25 April 2011.

Simon, Steven L., André Bouville, and Charles E. Land. "Fallout from Nuclear Weapons Tests and Cancer Risks." *American Scientist* 94 (January–February 2006). http://www.americanscientist.org/issues/feature/2006/1/fallout-from -nuclear-weapons-tests-and-cancer-risks. Accessed 17 January 2007.

Solnit, Rebecca. *Savage Dreams: A Journey into the Hidden Wars of the American West*. Berkeley: University of California Press, 1994.

Squitieri, Tom. "Bush Pushes for New Nukes." *USA Today*, 7 July 2003. http:// usatoday30.usatoday.com/news/washington/2003-07-06-nuclear-usat_x .htm. Accessed 3 April 2014.

Stahl, Sandra Dolby. *Literary Folkloristics and the Personal Narrative*. Bloomington: Indiana University Press, 1989.

———. "The Personal Narrative as Folklore." Master's thesis, Indiana University, 1975.

Stegner, Wallace. *The American West as Living Space*. Ann Arbor: University of Michigan Press, 1987.

———. *Mormon Country*, 3d ed. New York: Duell, Sloan & Pearce, 1942; Lincoln: University of Nebraska Press, 2003.

Sternglass, E. J. "Cancer: Relation of Prenatal Radiation to Development of the Disease in Childhood." *Science* 140 (7 June 1963): 1102–4.

Sulzberger, A. G., and Matthew L. Wald. "Flooding Brings Worry over Two Nuclear Plants." *New York Times*, 20 June 2011. http://www.nytimes.com /2011/06/21/us/21flood.html. Accessed 25 June 2011.

Szasz, Ferenc. "Downwind from the Bomb: A Review Essay." *Nevada Historical Society Quarterly* 30, no. 3 (1987): 182–87.

Szasz, Margaret Connell, ed. *Between Indian and White Worlds: The Cultural Broker*. Norman: University of Oklahoma Press, 1994.

Tabuchi, Hiroko. "Angry Parents in Japan Confront Government over Radiation Levels." *New York Times*, 25 May 2011. http://www.nytimes.com/2011/05/26 /world/asia/26japan.html?_r=1&scp=4&sq=japan&st=cse. Accessed 21 June 2011.

Takaki, Ronald. *Hiroshima: Why America Dropped the Atomic Bomb*. Boston: Little, Brown, 1995.

Thirty Years to Justice. Produced and directed by Ken Verdoia. KUED, 1985. Videocassette.

Thompson, Stith. *The Motif-Index of Folk Literature*. 6 vols. Bloomington: Indiana University Press, 1955–58.

Titus, A. Constandina. *Bombs in the Backyard: Atomic Testing and American Politics*. 2nd ed. Reno: University of Nevada Press, 2001.

Toelken, Barre. *The Anguish of Snails: Native American Folklore in the West*. Logan: Utah State University Press, 2003.

———. *The Dynamics of Folklore*. Rev. and expanded ed. Logan: Utah State University Press, 1996.

———. "Folklore and Reality in the American West." In *Sense of Place: American Regional Cultures*, edited by Barbara Allen and Thomas J. Schlereth, 14–27. Lexington: University Press of Kentucky, 1990.

———. "Traditional Water Narratives in Utah." *Western Folklore* 50 (April 1991): 191–200.

Udall, Stewart. *The Myths of August: A Personal Exploration of Our Tragic Cold War Affair with the Atom*. New Brunswick NJ: Rutgers University Press, 1994.

————. "We Aimed for Russia and Hit the West." *High Country News*, 22 August 1994. http://www.hcn.org/issues/14/438. Accessed 1 June 2011.

Union of Concerned Scientists. "Complex 2030: DOE's Misguided Plan to Rebuild the U.S. Nuclear Weapons Complex." December 2006. http://www.ucsusa.org/nuclear_weapons_and_global_security/nuclear_weapons/policy_issues/complex-2030-does-misguided.html. Accessed 6 December 2012.

United States Department of Justice. "Civil Division Compensation Programs: Radiation Exposure Compensation Act (RECA)." 7 June 2004. http://www.usdoj.gov/civil/torts/const/reca/about.htm. Accessed 2 September 2008.

University of Arizona Library. "Stewart L. Udall: Biographical Note," August 1997. http://www.library.arizona.edu/exhibits/sludall/biography.htm. Accessed 14 July 2009.

U.S. Atomic Energy Commission (AEC). "Atomic Tests in Nevada." Pamphlets, January 1955, 1957.

U.S. Atomic Energy Commission, U.S. Air Force, and Lookout Mountain Laboratory. *Atomic Tests in Nevada: The Story of AEC's Continental Proving Ground*. 1955. Videocassette. Private collection of Ken Verdoia, Salt Lake City.

U.S. Department of Energy (DOE). "History," n.d. http://www.energy.gov/about/history.htm. Accessed 20 October 2008.

————. *Human Radiation Experiments Associated with the U.S. Department of Energy and Its Predecessors*. Washington DC: Office of Human Radiation Experiments, 1995. http://www.rst2.edu/ties/radon/ramfordu/pdffiles/med8.pdf. Accessed 3 April 2014.

U.S. House of Representatives. *Health Effects of Low-Level Radiation: Joint Hearing Before the Subcommittee on Oversight and Investigations of the Committee on Interstate and Foreign Commerce, House of Representatives and the Health and Scientific Research Subcommittee of the Labor and Human Resources Committee and the Committee on the Judiciary, United States Senate, 96th Cong., 1st Sess., 19 April 1979* (Washington DC: U.S. Government Printing Office, 1979).

U.S. House of Representatives, Committee on Interstate and Foreign Commerce. *"The Forgotten Guinea Pigs": A Report on Health Effects of Low-Level Radiation Sustained as a Result of the Nuclear Weapons Testing Program Conducted by the United States Government*. Washington DC: U.S. Government Printing Office, 1980.

U.S. Senate. *Health Risks Associated with Atomic Tests in the 1950s: Hearing of Labor, Health and Human Services, and Education Subcommittee of Senate Appropriations, 1 October 1997* (Washington DC: U.S. Government Printing Office, 1997).

Utah Shakespeare Festival. "About Us: History." http://www.bard.org/about
/history.html. Accessed 14 September 2012.

Utah State University. "About Us." http://extension.usu.edu/about/. Accessed
14 March 2006.

Vansina, Jan. *Oral Tradition as History*. Madison: University of Wisconsin
Press, 1985.

Vyner, Henry. "The Psychological Effects of Ionizing Radiation." *Culture, Medicine, and Psychiatry* 7 (September 1983): 241–61.

Waconda, Alveno. Testimony. World Uranium Hearings, Salzburg, Austria, 16
September 1992. http://www.ratical.org/radiation/WorldUraniumHearing
/AlvenoWaconda.html. Accessed 18 October 2008.

Wagoner, Joseph K.; Victor E. Archer; Frank E. Lundin, Jr.; Duncan A. Holaday;
and J. William Lloyd. "Radiation as the Cause of Lung Cancer among Uranium Miners." *New England Journal of Medicine* 273 (22 July 1965): 181–88.

Walker, Ronald Warren, Richard E. Turley, and Glen M. Leonard. *Massacre
at Mountain Meadows: An American Tragedy*. Oxford: Oxford University
Press, 2008.

Ward, Chip. *Canaries on the Rim: Living Downwind in the West*. London: Verso,
1999.

———. "Fireworks Deferred: Divine Strake, Hellish Repercussions." TomDispatch.com, 29 June 2006. http://www.tomdispatch.com/post/97350/chip
_ward_on_pentagon_fireworks. Accessed 14 October 2008.

Wasserman, Harvey, and Norman Solomon. *Killing Our Own: The Disaster of
America's Experience with Atomic Radiation*. New York: Delacorte, 1982.

Watson, Bruce. "We Couldn't Run, So We Hoped We Could Hide." *Smithsonian*, April 1994, 47–57.

Weigle, Marta, and Peter White. *The Lore of New Mexico*. 1988; abridged ed.,
Albuquerque NM: University of New Mexico Press, 2003.

Weisiger, Marsha. *Dreaming of Sheep in Navajo Country*. Seattle: University of
Washington Press 2009.

Wellock, Thomas. "Radical Initiatives and Moderate Alternatives: California's
1976 Nuclear Safeguards Initiative." In Hevly and Findlay, *The Atomic
West*, 200–35.

White, Gordon Eliot. "AEC Put Lid on Leukemia Study." *Deseret (Salt Lake City)
News*, 29 January 1979.

———. "Deaths High in Utah Fallout Area." *Deseret (Salt Lake City) News*, 12
August 1977.

————. "Never Lost a Fallout Suit, AEC Said." *Deseret (Salt Lake City) News,* 15 February 1979.

White, Richard. *Remembering Ahanagran: Storytelling in a Family's Past.* New York: Hill and Wang, 1998.

Williams, Terry Tempest. *Refuge: An Unnatural History of Family and Place.* New York: Vintage Books, 1991.

Wills, John. *Conservation Fallout: Nuclear Protest at Diablo Canyon.* Reno: University of Nevada Press, 2006.

World Nuclear Association. "In Situ Leach (ISL) Mining of Uranium." March 2008. http://www.world-nuclear.org/info/Nuclear-Fuel-Cycle/Mining-of-Uranium/In-Situ-Leach-Mining-of-Uranium/. Accessed 5 April 2014.

World Uranium Hearings. "Poison Fire, Sacred Earth: Testimonies, Lectures, Conclusions—the World Uranium Hearing." Salzburg, Austria, 1992. http://www.ratical.org/radiation/WorldUraniumHearing/. Accessed 18 October 2008.

Worster, Donald. *Rivers of Empire: Water, Aridity, and the Growth of the American West.* Oxford: Oxford University Press, 1985.

Yazzie-Lewis, Esther, and Jim Zion. "Leetso, the Powerful Yellow Monster." In Brugge, Benally, and Yazzie-Lewis, *Navajo People,* 1–10.

Yeoman, Barry. "Unhappy Meals." *Mother Jones,* January/February 2003. http://www.motherjones.com/politics/2003/01/unhappy-meals. Accessed 23 August 2012.

Yih, Katherine, Albert Donnay, Annalee Yassi, A. James Ruttenber, and Scott Saleska. "Uranium Mining and Milling for Military Purposes." In Makhijani, Hu, and Yih, *Nuclear Wastelands,* 105–68.

Zeman, Scott C., and Michael A. Amundson, eds. *Atomic Culture: How We Learned to Stop Worrying and Love the Bomb.* Boulder: University Press of Colorado, 2004.

Zipes, J. *Breaking the Magic Spell: Radical Theories of Folk and Fairy Tales.* Austin: University of Texas Press, 1979.

INDEX

Page numbers in italics refer to illustrations.

burns, 69; beta, 87, 88
Burtenshaw, Ray, 67–68
Bushnell, Dan, 79, 192, 193
Bustad, Leo, 75

Califano, Joseph, 184
cancer, 145, 146, 151, 154, 155, 156,
    157, 175, 179, 181, 183, 219, 223;
    activism, 225; and atmospheric
    testing, 99, 102–4, 123, 138,
    141; detection, 200; and direct
    radiation exposure, 143–44;
    as impetus for action, 146–51;
    increase in, 183–85, 196, 210,
    228, 247n36; and radon, 9–10;
    skin, 83, 204; stomach, 42, 103;
    study, 184; treatment, 200, 212
Cane Valley AZ, 24, 44
Carrell, Carl, 104
Carrell, June, 103
Carter, Robert, 142–43
cattle, 67–69, 72–73, 136, 155; dairy,
    108; diet, 102, 106–7; Trinity, 66
Cedar City UT, 24; impact of radia-
    tion in, 65, 69, 82, 88–89, 117;
    LDS settlers and, 56; sheep
    industry, 55, 57, 61–62, 64, 72,
    74, 76, 83, 110, 170, 185
Cesium-137 (Cs-137), 105, 243n22
Charley, Perry, 86
Chenowith, William, 25
Chernobyl: power plant, 199, 212;
    Ukraine, 126
Christensen, Sherman, 81
Church of Jesus Christ of Latter Day
    Saints (LDS). *See* Mormon
Church Rock NM, 47
Church Rock Mill NM, 47–48

Church Rock Uranium Monitoring
    Project, 42
Citizens Call, 200–201
civil: defense, 4, 77, 143; drills, 128–
    29; health, 211; liberties, 195, 215
Clark, Doug, 69–72
cluster disease, 103, 133, 149–51, 160,
    161, 183, 188, 204, 220
Cold War, *152*; casualties, 226; colo-
    nialism, 119; and downwinders,
    222; policies, 149; pollution, 84;
    radiation, 10, 11, 129
collective memory, 220
colonialism, 119
Colorado Plateau, 21–23, 35, 48, 51, 121
Committee of Survivors, 189–90
commodities, 92, 110, 128
compensation, 101, 201, 204, 212–13,
    227; by AEC, 72
Comprehensive Nuclear Test Ban
    Treaty, 212
Confino, Alon, 228
Conner, Timothy, 211
Cooper, Ena, 133
Cooper, Evan, 156
Cooper Nuclear Station, 227
Corry, Annie, 50, 76, 83, 89, 188
Corry, Douglas, 73
cosmology of loss, 161
Cove AZ, 21, 30–31, 41, 90, 142, 153,
    179
Cox, Rulon "Boots," 112
Coyote Pass, 57
Crandall, Minette, 136
Crandall, Robert, 136
Crapo, Mike, 205
critical mass, 169–70; social, 170;
    supercritical, 169, 171, 212

crops, 94, 107, 123; failed, 94–95.
See also agriculture

dairy, 108–9, 115; industry, 95, 104,
108; promotion, 109–10. *See also*
milk
Dare, Wilbert, 31
David, Wilfred: "An Interim Report
of a Health Study of the Uranium Mines and Mills" by, 36
Davis, Devra, 199, 225–26
Dawson, Susan, 35, 39, 43, 205;
"Navajo Uranium Workers and
the Effects of Occupational Illness" by, 40–41
Dean, Gordon, 52
Deseret News, 65, 101, 133, 163, 169,
173, 174, 183, 185
Dickson DeBirk, Anne, 159
Dickson, Mary, 93, 112, 140–41, 158–
59, 159, 168, 191, 219, 220
discretionary function exemption,
193–94, 195
Divine Strake, 215, 218–19, 227
Down's Syndrome, 134, 146, 183
downwinders: acknowledgment
by government, 210; activists,
159, 199; compensable diseases,
203; defined, 202–3; discredited,
163; exposure, 158; illness, 122,
142, 146–47, 188, 210; lawsuit, 13,
76, 81, 192; as local color, 223;
region, 149–51, 161–63, 172, 183,
185; subsequent generations,
209–10, 220; testimony, 186–88.
See also narrative; uranium-
affected people

Doyle, Henry: "An Interim Report of
a Health Study of the Uranium
Mines and Mills" by, 36
Duchesne County UT, 95, 108, 115
Dugway Proving Ground, 68, 85,
161
Durango CO, 30–31
dust, 48, 123; concentration of, 39;
exposure to, 39; in mine, 37;
poison, 4; precautions, 39–40;
radioactive, 15, 42, 61, 73, 122

Eichstaedt, Peter, 25, 37, 44
Elko NV, 84, 135
employment, 24–25, 26–28, 30, 34,
216, 217; economic loss, 8, 200;
economic solvency, 167; and
money, 181–82; wage labor, 2,
22–23, 25, 27, 31, 33, 48
Energy Fuels Company, 217
Energy Research Foundation, 211
environmental damage, 149, 171–72,
208, 216, 225
environmental justice, 189–90, 208
Environmental Protection Agency
(EPA), 45, 46
epidemiology, 11–12, 184, 197; health
risks from radiation exposure,
194; and lifestyle factors, 187–88;
popular, 147–49; studies, 195,
201, 204, 211, 227
"Estimated Exposures and Thyroid
Doses Received by the American
People from Iodine-131 in Fallout" (National Cancer Institute),
210
Evans, Delayne, 70, 76

experiential knowledge, 73–74, 89, 123, 139–40, 170, 223
Exposed (play), 160

Facts about Fallout (Federal Civil Defense Administration), 126
fallout: affecting livestock, 59, 61, 64, 68; calculator, 210; concentrated, 67, 105, 150; distribution, 16, 53, 60, 133, 136; exposed areas, 159, 197, 201–3; 247n34; and foodchain, 74, 108, 112, 113, 156; monitoring, 13, 113, 202; related illness, 10, 104, 106, 146, 157, 173, 197
farming, 8, 21, 28, 32, 48, 55–56, 96, 99, 104. *See also* agriculture
Federal Civil Defense Administration (FCDA), 126, 129; *Facts about Fallout* by, 126
Federal Tort Claims Act, 193
Federation of American Scientists, 9
First Mesa AZ, 96
Fischer, Frank, 14, 192, 212
Fisher, George, 104
Fisher, Susan, 104
folklore, 17–19, 29, 122–23, 134
folklorist, 17, 58, 187
food, 154, 160; chain, 8–9, 12, 16, 39, 44, 48, 107–8, 197; contamination, 40, 101, 104, 108, 112–13, 115, 121, 155–56, 220; local, 171, 181, 182, 188; production, 220–21, 223; traditional, 96, 101, 122
foodshed, 90; radiation in, 93, 122, 187

Ford, Bacon, and Davis Utah, Inc., 45
Fort Calhoun Nuclear Station, 227
Four Corners region, 2, 7, 24, 204; abandoned sites, 200; residents, 45, 48; and uranium, 22–23, 31
Fraga, Gilbert, 87–88
Fukushima Daiichi Nuclear Power Plant, 227
funding, 10, 106, 195–96, 197, 198, 199, 211, 227

Gallagher, Carole, 158
Gallup NM, 47, 48
Geertz, Clifford, 18, 73
Geiger counter, 58, 63, 64, 66, 68, 115, *116*, 117–18, 136, 155–56, 157–58
General Electric Company, 75
Gofman, John, 9, 184
Gordon, Janet, 200
Goulding, Harry, 24, 28–29
government: assurances, 130, 171, 184, 215; belief in, 60, 73; betrayal by, 8–9, 15, 33, 89, 155, 170–71, 188, 207, 208; compensation by, 159, 180, 195, 204; cover-up, 68, 82, 185, 186, 199; deception, 8, 13, 56, 75, 78, 79, 132, 144, 160, 185, 186, 188, 193; dismissal by, 70–71, 147; dissent, 7, 77, 114, 224; integrity, 171–72; loyalty to, 60–61, 77, mistrust of, 146, 212; official fiction, 225; public relations, 128, 130; remediation, 149, 171; withholding warnings, 137, 188, 194
Granddaddy A bomb, *164*
Grants Mineral Belt, 25

Three Mile Island: accident at Nuclear Power Plant, 47, 126, 187, 199; Nuclear Generating Facility, 185, 187

Tikaboo Valley NV, 54, 57, 58

Timothy, Dave (DT), 95, 97, 101, 104, 109, 112, 115, 122–23; as plaintiff, 198

Timothy, James (father of DT), 95–96

Timothy, LaWanna (mother of DT), 96

Tome, Harry, 121, 141, 175, 178, 201

Tonopah NV, 101, 135–36

Tonopah Times-Bonanaza, 135–36, 164

Torres, Israel, 143

Treaty of Ruby Valley, 52

Trinity cattle. *See* cattle

Trinity site NM, 223

Trum, Bernard, 72–73

Truman, Harry, 109–10, 213

Truman, Preston Jay, *116*, *130*, 201, 204

Tsosie, Minnie, 153, 175

Tuba City NM, 42, 46

Tutt, George, 2, 20, 28, 37

Twin Springs NV, 133

Udall, Stewart, 176–79, 192, 194

Udall, Tom, 205

Uintah Mountains, 95, 96, 107, 108

United Nuclear Corporation, 47–48

University of New Mexico Family and Community Medicine Department, 208

uranium: boom, 27, 115, 123, 161, 162, 180, 182, 188, 217; exposure, 154, 207; extraction, 27, 43, 118, 216, 219, 224, 228; families, 15, 180, 194, 224; ionization by body, 167; lease, 26–27, 32; mines, 2, 24, 25, 87, 129, 153, 194–95; mine waste, 44, 94, 123; ore, 43–44; prospectors, 115–17; safety standards, 167; scavenging, 34; tailings piles, 44–45, 94, 200, 220; tailings spill, 47–48; transportation, 41, 42, 48; tribal opposition to, 216, 218; widows, 175, 178–79, 189. *See also* activists; mine waste; radiation: related illness

uranium-affected people, 212, 221, 222, 227, 228; as "local color," 223; marginalization of, 223. *See also* downwinders

uranium industry, 25, 29, 30–33, 35, 36, 141, 158, 167; decline, 199, 213, 216, 220, 227; exploitation by, 119, 170, 182; sites, 200; survivors, 201

Uranium Mill Tailings Remedial Action Project, 200

Uranium Radiation Victims Committee, 201

U.S. Department of Defense (DOD), 196, 203, 215

U.S. Department of Energy (DOE), 143, 175, 186, 196, 203, 215–16, 219; Energy Research and Development Administration, 45, 186, 198; Nuclear Regulatory Commission, 186, 198; *Human Radiation Experiments Associated with the United States Department*